T0122994

Life Histories of Genetic Disease

Life Histories of Genetic Disease

Patterns and Prevention in Postwar Medical Genetics

Andrew J. Hogan

JOHNS HOPKINS UNIVERSITY PRESS BALTIMORE

Printed in the United States of America on acid-free paper
9 8 7 6 5 4 3 2 1

Johns Hopkins University Press
2715 North Charles Street
Baltimore, Maryland 21218-4363
www.press.jhu.edu

Library of Congress Cataloging-in-Publication Data

Names: Hogan, Andrew J., author.
Title: Life histories of genetic disease : patterns and prevention in postwar medical genetics / Andrew J. Hogan.
Description: Baltimore : Johns Hopkins University Press, 2016. | Includes bibliographical references and index.
Identifiers: LCCN 2015050732| ISBN 9781421420745 (hardcover : alk. paper) | ISBN 1421420740 (hardcover : alk. paper) | ISBN 9781421420752 (electronic) | ISBN 1421420759 (electronic)
Subjects: | MESH: Genetic Diseases, Inborn—prevention & control | Genetics, Medical—history | History, 20th Century | History, 21st Century
Classification: LCC RB155 | NLM QZ 11.1 | DDC 616/.042—dc23
LC record available at http://lccn.loc.gov/2015050732

A catalog record for this book is available from the British Library.

To my parents
Pat and Gerry

Contents

Preface

When I began this project in 2009, I intended to study the creation, promotion, and uptake of new prenatal diagnostic approaches. I interviewed the developers and early medical adopters of these technologies and examined why new techniques, such as chorionic villus sampling (CVS), which many believed to be superior, did not successfully outcompete amniocentesis, the existing gold standard. After a few months of considering the history and practices of prenatal diagnosis, however, I became aware that sampling fetal cells and detecting disease-causing mutations among them was just one aspect of a much broader story. Equally significant in this history were questions of how and why physicians sought to identify genetically distinct disorders, as well as how geneticists established one-to-one associations between clinical conditions and genetic mutations. I was interested in how physicians and geneticists developed the confidence necessary to diagnose a disorder based on a mutation that was made visible prenatally, with few or no clinical findings to back it up. This struck me as a substantial consideration, given that a diagnosis often led parents to choose preventive abortion.

Aware of the wide array of disorders that could be diagnosed prenatally, nearly a decade after the completion of the Human Genome Project, I decided to conduct a series of in-depth historical investigations of how medical geneticists delineated genetic disorders, correlated them with genetic mutations, and worked to make these conditions detectable and preventable prenatally. The conditions I chose to study were rare in comparison to the major causes of morbidity and mortality in society but still quite common among discrete genetic disorders. Familiarity played a role in my selection. One of the diseases affected my family, another was an exemplar in undergraduate genetics courses, and the third was closely tied to Philadelphia, where I was pursuing my doctoral degree. Importantly, I also knew that the history of each disorder would reflect the great promises of genetic medicine during the late twentieth century and the frustrations involved in establishing a reliable genetic marker for presymptomatic diagnosis.

Molecular testing now exists for the three disorders that I examine, but the history of genetic diagnosis for each began with the microscopic examination of chromosomes. The history of mapping human genes and diseases has been remembered and presented largely in the context of advances in molecular biology. However, gene and disease mapping began in the early 1970s with the development of new techniques for visualizing and manipulating human chromosomes. This book examines how human chromosomes came to embody genetic diseases. The analysis of chromosomes was central to the practices of medical genetics from the 1950s into the early-twenty-first-century era of whole genome molecular screening. Throughout this period, medical geneticists situated human chromosomes as their primary organizational units as they developed an infrastructure for diagnostic interpretation, analysis, and communication. The following chapters explore the representations, continuous growth, and evolving use of this infrastructure.

Infrastructures are embedded all around us and play a central, if often overlooked, role in our everyday lives. Within a community, infrastructures function to help people achieve a variety of daily tasks. They shape how we move about the world, our expectations about what can reasonably be accomplished, and how we organize knowledge. When we think of infrastructure, we often consider the large-scale artifacts and systems that we rely upon daily, including roads and bridges, subways, sewers, and electrical grids. These systems were built to accommodate many uses and end goals, some of which were easy to imagine ahead of time, others that were never foreseen, and a few that were considered inappropriate. Highways were not built to facilitate drug trafficking, nor subways for picking pockets, but they were inevitably put to these ends. The mixed use of infrastructure is central to its value and is also the basis for its regulation.

Professions also rely on infrastructures for the organization of knowledge and completion of daily tasks. Use of these infrastructures is much more limited than a city subway network is, but they still play a significant role in shaping the thinking and practice of many individuals. Indeed, developing a working knowledge of an infrastructure is often central to gaining membership in a profession. I focus here on the development of an infrastructure for organizing genetic information. Like other infrastructures, this one was built on a set of standards and was put to use by a wide array of practitioners from different specialties; it is seamlessly embedded in many locations throughout the discipline of medical genetics. Medical geneticists become familiar with this infrastructure early in their training, come across representations of it and references to it on a regular basis, and retain an awareness of its basic makeup in their memory, to aid in the interpretation and communication of findings. I demonstrate that the chromosomal infrastructure of genetic medicine had a significant presence in the field from the 1970s and was integrated as a central component of laboratory and clinical workspaces, online databases, and textbooks.

Infrastructures have both foreseen and unintended consequences. Contributions to the chromosomal infrastructure most often came from studies in which researchers were focused on improving the diagnosis, understanding, treatment, and prevention of a specific disorder. Over the years, this research added to the understanding of many genetic disorders and in some cases improved their treatment. Frequently, results of this research were also put to use in facilitating prevention. Taken as a whole, the chromosomal infrastructure provided medical geneticists with information about the likely clinical implications of randomly identified mutations. After 2000, with this infrastructure in place for reference and a new set of whole genomic screening platforms available, medical geneticists and biotechnology entrepreneurs expanded the scope of genetic testing toward the identification of any potentially disease-causing mutation. In the prenatal context, with few or no bodily signs of disease available to supplement genetic findings, medical geneticists relied on information from the chromosomal infrastructure to provide parents with a diagnosis and in some cases the option of preventive abortion. Findings that researchers had previously reported as part of studies on specific disorders thus became part of a larger regime for prevention before birth. In this book, I examine the construction of the life histories of genetic diseases, including their association with a mutation, and offer a historical account of the diffuse collection of studies and aims in medical genetics that contributed to the piecemeal development of an infrastructure for genetic diagnosis and (ultimately) prenatal prevention.

Acknowledgments

Many people and institutions have contributed over the past eight years to making this book possible. The University of Pennsylvania provided generous support through the Benjamin Franklin Fellowship, the George L. Harrison Graduate Fellowship, and a Dissertation Research Fellowship. The Department of History and Sociology of Science offered a supportive and intellectually stimulating scholarly home for me for five years. It is a department that provides unusually generous support and resources for its graduate students, and I am very grateful for the years I spent there. I also benefited significantly from the support and enthusiasm of my colleagues at the University of Virginia and Creighton University. These institutions provided the financial and scholarly support and, most importantly, the time I needed to complete this project, as I began my career as a teacher and scholar of history.

Many individuals have contributed to shaping this book project over the years. I am indebted to my dissertation adviser, Susan Lindee, who helped me make the initial connections with researchers in Philadelphia that got my research project going. Susan has a great skill for providing helpful support and reassurance when it is needed, while asking the tough questions when they are necessary. Ruth Schwartz Cowan also played a significant role in shaping how I approached my research, especially on prenatal diagnostic technologies. Importantly, Ruth also enjoys helping researchers to make social connections, and she helped me find my way in the medical community at the University of Pennsylvania. John Tresch, Jonathan Moreno, and Robert Aronowitz were also constant sources of advice and support. Their doors were always opened to me when I needed someone to talk to about this project and my career more broadly.

Many others at the University of Pennsylvania also gave significant support during the years of research for this project. In particular, I want to thank the HSSC graduate student community, past and present, for all that they contributed to the project and to my scholarly and social life. Many of my fellow Penn alumni, including Whitney Laemmli, Sam Muka,

Rachel Elder, Joanna Radin, Kristoffer Whitney, Jessica Mozersky, Mary Mitchell, Matt Hoffarth, Sam Beckelheymer, Doug Hanley, and Andy Fenelon, helped to make my years in Philadelphia some of the most fun and productive of my life. At Penn, I also benefited greatly from my membership in the Center for the Integration of Genetic Healthcare Technologies (CIGHT). This organization helped me to engage with a broad network of biomedical professionals who were crucial to the completion of this project. Reed Pyeritz, Barbara Bernhardt, Michael Mennuti, and Laird Jackson provided significant support and assistance, while making it clear that the results of my project were of interest and value to the broader medical community. I was very fortunate to have been at Penn during the years when CIGHT was best funded and most active. It is hard to imagine completing my book without this social network.

The history of science and medicine community has been a constant source of energy and encouragement. I have been lucky to get to know many of the wonderful members of this community from across the world. Over the years, I have benefited significantly from the friendship, feedback, and career advice offered by many scholars, including Nathan Crowe, Dawn Digrius, Luis Campos, Angela Creager, Soraya de Chadarevian, Stephen Pemberton, Henry Cowles, Courtney Thompson, Jenna Healey, and Nathaniel Comfort. In particular, I want to thank Robin Scheffler and Stephen Casper for their feedback and advice on this manuscript. They were instrumental in helping to get the book into shape, after having read my entire first draft. I also want to thank my editor, Jackie Wehmueller, at Johns Hopkins University Press, for all of her help, enthusiasm, and support in bringing this book project to fruition.

I owe a debt of gratitude to those who have agreed to be interviewed or have helped to provide resources for this project: Phoebe Letocha and Andrew Harrison at the Chesney Medical Archives, David Rose at the March of Dimes Archives, and Uta Francke at Stanford University. More than 30 geneticists and clinicians have been kind enough to take the time to be interviewed, and the project certainly would not have been nearly so successful without them. Together, they have provided a valuable data set that was integral to my research. I want to thank all of them for their time, energy, trust, and interest in the project.

Most importantly, I thank members of my family, without whom I would never have gotten this far. My parents have provided endless love and support for many decades. My father, Gerry Hogan, imparted to me a love of history, and my mother, Pat Hogan, a continuous sense that both the past and the future of genetics is an important topic of interest and study. Also, I thank my sister Lauren, who provided much support and valuable insights based on her own growing interest in genetic medicine. Finally, I thank, with great affection, my wife, Sabrina Danielsen, who entered my life just before my research for this book began. Sabrina always offered unwavering love, support, and belief in me and in the project. She helped me to see and analyze the world in ways that will forever enhance my life and scholarship.

Life Histories of Genetic Disease

Introduction

Pursuing a Better Birth

When embarking on a new pregnancy, parents face a daunting array of risks and choices. Among decisions about diets, birth plans, parental leave, car seats, and child care is the question of whether to undergo prenatal testing and, if so, what amount and kinds of genetic information to receive about the fetus. Throughout the postwar period, the scope of choices for prenatal testing varied across national contexts. The United States followed a free market approach, while many socialized health care systems in Europe offered fewer options. Though the range of choices varied, beginning in the late 1970s women felt increasing social and medical pressure to undergo some form of prenatal screening. Even if women were offered the choice of "no," in choosing to decline, they risked negative social and medical judgments.[1] As prenatal diagnosis became more common in the late twentieth century, it was primarily targeted to identifying specific disorders in certain populations. Because of an increased risk for chromosomal abnormalities with advanced maternal age, physicians, public health officials, and medical geneticists encouraged pregnant women over age 35 to pursue prenatal testing for Down syndrome. During the 1980s and 1990s, medical geneticists also developed new prenatal testing options for disorders known to run in families, such as Tay-Sachs disease, thalassemia, sickle-cell anemia, Huntington's disease, and cystic fibrosis.

The twenty-first century has seen a significant expansion in what could be tested for prenatally. This growth was driven largely by diagnostic laboratories at research universities and biotechnology startups in the private sector, through the development of increasingly powerful and dense testing platforms. Rather than choosing to test only for certain disorders based on parental or familial risk, parents often found it more cost effective to test for tens or hundreds of disease-specific mutations at once. After 2005, medical geneticists and biotechnology companies also began offering whole genome prenatal screening, pushing testing further toward the search for any disease-causing mutation. While the density of genetic findings had increased significantly, the results of whole genome screening would have been of little value without resources for interpreting their clinical implications. This book is about the piecemeal development, since 1970, of an infrastructure for genetic diagnosis. During the first two decades of the twenty-first century, medical geneticists rely on this infrastructure to link randomly identified mutations to specific disorders, and in doing so, they are greatly expanding the scope of prenatal diagnosis and prevention.[2]

Eugenics and Medical Genetics

The desire for a better birth has long animated science, medicine, and social policy, as well as the decisions of parents. This goal is the basis of eugenics, a term meaning "wellborn" coined by Francis Galton in 1883. Ever since Galton, eugenic aims have been pursued by various means on both an individual and a societal level. During much of the twentieth century, eugenic ambitions tended toward the large scale, with a focus on improving the health and purity of society, more than improving the well-being of any individual. Participants in the American Eugenics Movement of the early twentieth century were fixated on the type of people that were reproducing. They wanted to see more people like themselves—white, middle- or upper-class—bearing children, and fewer immigrants, racial minorities, and impoverished families doing so. Eugenic social interventions, including government sanctioned sterilization in most US states, aimed to reduce the incidence of feeblemindedness, epilepsy, deafness, and alcoholism; but primarily it focused on limiting the reproduction of populations who were prejudicially assumed to possess these genetic traits.[3]

Many geneticists and physicians regarded the aims and tactics of early-twentieth-century US eugenicists as overly simplistic, misinformed, and ineffectual. This is not to say, however, that these practitioners did not believe in the potential for genetic research to contribute to public health. The earliest medical geneticists viewed the identification and removal of disease-causing genes from society as an important public health effort that would aid both affected families and society at large. The practice of medical genetics in the United States had its origins in the heredity clinics of the 1940s and 1950s. Heredity clinics were housed at major research universities and staffed by partnerships between physicians interested in genetics and geneticists in medicine. These practitioners primarily saw individuals who were interested in the heritability of disorders that affected their families. Medical geneticists of this era drew family pedigrees to trace the inheritance pattern of a particular trait and, from this information, to predict its likelihood for reoccurrence. Similar practices were central to genetic counseling into the early twenty-first century.[4]

Heredity clinics and the field of medical genetics were institutionalized during the decades around World War II, when Nazi atrocities revealed to the world the length to which eugenic ideals could be taken. During the 1930s, German sterilization programs, modeled on legislation first passed in US states, had been the envy of American eugenicists. The Holocaust undoubtedly chastised eugenicists, but it did not directly lead to the changing of policy or practice. Eugenic sterilization was ongoing in many US states into the 1970s; in some localities it even increased in prevalence after World War II. Heredity clinics continued to provide genetic consultations, which were still tinged by racial assumptions and desires to improve society at large through the eradication of defective traits. The field's first professional organization, the American Society of Human Genetics, was led into the 1960s by medical geneticists who continued to support socially oriented eugenic views. The turn toward autonomy and choice in genetic medicine, putting the interests and desires of individuals ahead of purported societal benefits, came in the 1970s.[5]

Eugenics was altered, but not eradicated, in the United States after 1970. States replaced top-down sterilization policies with financial and policy support for preconception and prenatal testing. Within this new framework, parents were offered a genetic risk assessment and given choices about how to proceed. In all but a few cases, prevention was the only option medical geneticists could provide, through partner selection, forgoing reproduction, or prenatal testing and targeted abortion of affected fetuses. Social critics questioned the larger implications of these testing regimes and noted the role of prenatal diagnosis in changing the experience of pregnancy and commodifying its results.[6] "Liberal eugenics," as some called these practices because of their free market availability, offered parents choices about which children they wanted to bring into the world. However, many scholars noted that parents' decisions were strongly biased by the economic interests of the state, corporate desires to sell new forms of testing, medical presumptions about the impact of disability, and the messages that they received about what types of children were acceptable to society at large.[7]

Life Histories of Genetic Disease examines the role of postwar medical genetics in facilitating and enhancing eugenic choice. The extent to which the medical genetics model of disease prevention resembled the aims of eugenicists before 1970 has remained a point of scientific, social, and scholarly debate and is a consideration that animates the focus and analysis of this book.

Subspecialties in Medical Genetics

Beginning in the 1950s, a new generation of medical geneticists sought to distance their approaches and aims from the term and practices of eugenics. Postwar medical geneticists maintained some of the techniques of their predecessors, including family pedigree studies, but also adopted approaches from many other specialties, in an attempt to improve the targeting of genetic disease and demonstrate the broad scientific and medical basis of their discipline. Among the various biomedical professionals who were drawn to medical genetics after 1960 and developed new subspecialties, human cytogeneticists and dysmorphologists introduced some of the most significant visual and analytic methods and tools used in postwar medical genetics.

Postwar medical genetics was home to significant professional diversity. Internal divides between physicians and geneticists shaped its early history in the United States. The first professional organization in the field was called the American Society for Human Genetics, a name that reflected the significant influence of scientists who studied human genetic variation. Over the next two decades, physicians increasingly entered the field, so that by the 1970s its makeup was evenly split between MDs and PhDs. James Neel, who held degrees in both areas, helped to lead the way in hybridizing medical genetics, as did physicians interested in the genetic basis of disease, such as Victor McKusick, Kurt Hirschhorn, and Arno Motulsky. Throughout the postwar period, these medical geneticists also identified themselves by additional designations: human geneticists if they worked primarily in a laboratory setting, and clinical geneticists when seeing patients was their main daily activity.[8]

There was also significant diversity among the physicians and the biologists who initially entered medical genetics. The first generation of physicians who got involved in medical genetics included internists (specialists in adult medicine), pediatricians, and obstetricians. In the 1970s some of these physicians also began to populate the new medical genetics subspecialty of dysmorphology, which focused on delineating and naming genetic syndromes. Many biological fields also contributed to medical genetics, including population and statistical (classical) genetics, mammalian genetics, cytogenetics (the study of chromosomes), biochemistry, and molecular biology.[9] This book focuses primarily on the contributions of dysmorphologists and human cytogeneticists to postwar medical genetics, especially how they made genetic disease visible and made sense of it and its causes, in the laboratory and the clinic, between the 1970s and the 2010s.

Pediatrician David W. Smith coined the term *dysmorphology* in 1966 to distinguish his approach of studying human malformation from that of teratology, a Greek term meaning "the study of monsters." Dysmorphology differed from human teratology in two major ways. First, it assumed that the primary causes of human malformation were genetic in origin, rather than environmental. Second, dysmorphology focused on studying patterns of bodily malformation, including both major and minor defects. Smith's focus on patterns was rooted in his belief that paying attention to multiple malformations would make it possible to delineate disorders that had a single genetic cause. Single major malformations such as cleft palate, clubfoot, or mental deficiency, were likely to be caused by many different mutations. But, Smith argued, if multiple malformations occurred together, or along with minor anomalies of the face, feet, or hands, it was likely that a single distinct and identifiable genetic mutation caused this clinical pattern.[10]

Smith introduced dysmorphology during a period when many physicians and teratologists were critical of single-gene explanations of bodily malformation. Prominent among them was Josef Warkany, who studied human congenital (inborn) malformations in animal models. He had grown up in Vienna, and while he came to the United States before the rise of Hitler, his rejection of genetic causes of malformation in favor of environmental explanations was likely a response to Nazi eugenics. To the extent that Warkany accepted genetic causes of malformation, he understood them to be polyfactoral rather than single mutations. Seeking a middle ground, F. Clarke Fraser, an early Canadian medical geneticist, brought genetics back into teratology during the 1950s with his own work on mice. He demonstrated that the same environmental mutagen caused different degrees of palate clefting in genetically distinct mice. Fraser's work showed that both genetic and environmental factors contributed to the severity of bodily malformation. He also introduced the concept of genetic heterogeneity, which notes that the same outcome, for example, cleft palate or diabetes, could result from different genetic mutations in different individuals.[11]

Breaking with his immediate predecessors, Smith approached the causes of congenital malformations from a more exclusively genetics-oriented perspective. In doing so, Smith fashioned dysmorphology as teratology for medical geneticists. Smith's method for delineating pediatric disorders was influenced by his early work at the University of Wisconsin, Madison, with cytogeneticist Klaus Patau during the late 1950s. Patau asked for Smith's help in identifying patients with multiple bodily malformations that might be suggestive of a major chromosomal abnormality. Plant cytogeneticists had previously shown that the gain or loss of entire chromosomes was associated with complex variations in phenotype. For decades, medical geneticists had pondered the possibility that similar abnormalities in humans might cause multifaceted disorders like Down syndrome. At Wisconsin, Smith identified multiple newborns who showed a pattern of congenital malformation suggesting a major chromosomal defect, and Patau reported finding an extra chromosome in the cells of a few of these patients. Some called the disorder Patau syndrome and later, after the specific chromosome was identified, trisomy 13.[12]

The collaborative approach of Smith and Patau was representative of medical genetics practice throughout the postwar period. Their work involved the combination of two distinct traditions, rooted in differing ways of making visible and understanding human variation. Importantly, Smith and Patau performed their visual analyses with the same presumption in mind about the nature of disease. Each researcher believed that one clinical disorder could be reliably associated with one chromosomal mutation. This perspective informed the development of both human cytogenetics and dysmorphology after 1960 and was central to the thinking and analysis of medical geneticists more broadly over the next half century as practitioners from these two specialties worked together to build postwar medical genetics.

Ways of Seeing in Medical Genetics

Dysmorphology was a fundamentally clinical discipline, practiced exclusively by physicians. It was no accident that it first emerged from pediatrics. In dysmorphology, some of the most interesting disorders also had the highest mortality, leading to death in early childhood. Infants born with trisomy 13 often died within weeks or months. For Smith, a primary goal of dysmorphology was to determine the mechanism by which congenital malformations came about and evolved throughout development, a focus that stretched back to conception and followed a patient forward into childhood and even adulthood. The clinical assessment of reoccurring developmental patterns of malformation was central to how dysmorphologists studied and understood genetic disorders.

Jon Aase, a student of Smith's, compared dysmorphology to detective work. This was a common trope among dysmorphologists and those who worked with them: dysmorphologists had a talent for noticing subtle features, which anyone else would miss, and attached significant meaning to them.[13] As part of their standard examination of patients, dysmorphologists looked for and measured features that most physicians would not consider relevant to disease. They paid attention to the distance between a patient's eyes, ears, and nipples, as well as the size and shape of the nose, forehead, cheeks, eyes, and ears. In addition, dysmorphologists looked for malformations of the fingers and toes, along with irregularities in finger, hand, foot, and toe prints. Each of these features, while extremely minor from a clinical perspective, was understood to be a potential component of a larger and reoccurring pattern of bodily malformation. As Smith put it, "Minor malformations, structural aberrations which are of little or no medical or psychologic consequence to the patient are frequently overlooked or disregarded as being of no significance. They may, however, represent significant clues. . . . In the diagnosis of a multiple malformation syndrome, minor anomalies may help in determining whether a major defect such as mental deficiency has its onset in prenatal life." Dysmorphologists looked for instances in which malformations such as cleft palate, heart defects, and abnormal facial features occurred as part of a larger developmental pattern. Their investigation of the body was, in the words of Aase, like Sherlock Holmes's investigation of a crime scene. By looking for certain patterns of bodily malformation, which they believed were probably caused by one initial event, dysmorphologists sought to piece together a story of what happened and trace it backward to identify a culprit. In most cases, they suspected a genetic mutation.[14]

The techniques for examining the human body that dysmorphologists drew upon were nothing new. Anthropometric measurements had been a standard part of physical anthropology dating back to the nineteenth century, as was physiognomy, the study of faces. These approaches had been in use for many years to define and quantify human difference. In past iterations, they had been used to reify the superiority of the white race, make sense of differences in intelligence, and identify criminal types.[15] Dysmorphologists also interpreted these traits as fundamental and irreversible reflections of human difference, which they believed were caused by genetic variants. However, rather than highlighting the traits as reasons to cast blame on parents for the faults of their children, dysmorphologists presented them as a means for exoneration. If they identified a pattern for malformation in a child that was believed to result from a genetic mutation, this implied that there was nothing a mother had done wrong or could have done differently to avoid the malformation, such as avoiding exposure to an environmental mutagen. Dysmorphologists assumed that the condition was caused by a random genetic mistake, meaning that no one was to blame.

In his writing, Smith pointed to the importance of providing parents with a clear explanation of their child's inborn bodily malformations, as well as any information that could be offered about the potential for reoccurrence in future pregnancies. He also carefully managed parents' interpretations of their child's condition by framing its origin as "a normal stage in development which failed to undergo completion." Smith chose to characterize genetic conditions in terms of defects and patterns of malformation, rather than as forms of abnormality. He advised that physicians, when counseling parents, isolate the malformations from the child, so as to "prevent the parents from considering the whole child abnormal on the basis of one structural defect." Smith preferred to focus on abnormal processes of development rather than abnormal individuals.[16]

In his extensive clinical search for complex and reoccurring patterns of malformation, Smith was pursuing a reductive approach to medicine. Dysmorphology was based on his belief that if a pattern of malformation was multifaceted and yet happened independently in multiple patients, then it must represent a real medical entity, with a single cause. While many physicians assumed that a child with multiple malformations was probably affected by a variety of genetic and environmental factors, dysmorphologists thought otherwise. Smith and his followers preferred explanations that focused on one developmental cause, often genetic, as the basis for all of the major and minor malformations identified in an individual or a group of patients. Smith tailored this reductive clinical approach to fit the presumptions of postwar medical genetics.[17]

Dysmorphology, like medical genetics more broadly, was an urban profession and an international pursuit. This was because most of the disorders that dysmorphologists studied were rare, occurring at a rate between 1 in 5,000 and 1 in 50,000 births. As a result, dysmorphologists gravitated toward population centers of a million or more. Without a high birth rate in their community, dysmorphologists were unlikely to see multiple cases of the same genetic disorder. While a dysmorphologist in the 1970s and 1980s could flip through a textbook to identify a disorder, the best way to develop familiarity and recognition was through exposure to multiple affected patients. Even within one geographic region, dysmorphology was practiced more by a community than by an individual. No one doctor could ever expect to see or memorize the majority of known patterns of malformation. Thus, central to dysmorphology throughout its development was the convening of local, regional, and international groups to compare cases and decipher disorders, at the bedside and based on photographs. Dysmorphology was a visual and analytic skill that one developed over a lifetime. It was not enough to just recognize patterns of malformation; one also had to learn to discern how a malformation differed from a list of similar alternative diagnoses.[18]

While dysmorphologists examined and measured human bodies in the clinic, drawing upon markers and measurements that were visible to the naked eye, the work of human cytogeneticists was largely dependent on microscopic observation. Unlike human bodies, chromosomes were not readily visible to the clinical gaze, even with sufficient microscopic magnification. Postwar human cytogeneticists relied on several biological, chemical, and mechanical processes, including cell reproduction and staining, to make chromosomes visible and analyzable. Chromosomes were not examined as they were found, but rather were rearranged into a karyotype by relative size. Cytogeneticists looked for any absent, extra, or rearranged genetic material, as well as other abnormal structural features. As a whole, the human chromosome set represented the entire genome for human cytogeneticists; any visual abnormality that was identified among the chromosomes was also understood to affect a discrete genomic location.

The approaches of postwar human cytogenetics varied among practitioners and laboratories. Reflecting these local idiosyncrasies, scholars have referred to human cytogenetics as an "artisanal" and "craft" practice that defied automation and standardization. Human cells and chromosomes often behaved in unanticipated ways, and local practices were regularly adjusted in response to days and weeks of poor results. Over the decades, computer programs were developed for use in distinguishing and interpreting chromosomes, but their adoption in the laboratory either failed or was incomplete. Human cytogenetics was also a more professionally diverse field than dysmorphology. While most laboratory directors held a PhD, the growing demand for prenatal diagnosis in the 1980s led to the hiring of many clinical cytogeneticists with no more than master's-level training. Cytogenetic analysis was a skill set that undoubtedly improved over a lifetime, but it was also one in which a technician could gain proficiency in a matter of months.[19]

The most significant and ongoing accomplishment for postwar human cytogeneticists in standardizing their craft could be seen in the ideograms that they created of each chromosome, beginning in 1960, and regularly updated every few years. Ideograms were meant to be idealized representations of a chromosome's distinguishing features. Every human cytogeneticist became intimately familiar with the set of standardized ideograms over his career. However, new cytogeneticists did not learn to recognize individual chromosomes by studying ideograms. Chromosome recognition and analysis could be learned only under the microscope, because ideograms were not meant to look like the chromosomes that cytogeneticists actually saw. Rather, ideograms were designed for the comparison and communication of findings.[20]

Chromosomes in the Clinic

Human cytogenetics was adopted as a clinical tool during the 1950s in response to concerns about the impact of increasing levels of nuclear radiation on the human gene pool. In a 1950 paper, geneticist Herman J. Muller pointed to the potential for radiation to greatly increase the number of disease-causing mutations in the human population. Geneticists were unsure how to best detect these mutations and track their occurrence following radiation exposure. The microscopic examination of chromosomes provided one potential approach. As historian Soraya de Chadarevian noted, chromosomal analysis "offered a glimpse of the complete genetic make up on an individual." In this era, medical geneticists began to think of the chromosomes as a visible embodiment of the human genome, among which potential disease-causing mutations could be identified. While the fear of nuclear fallout and its impacts decreased over subsequent decades, cytogenetic analysis became an increasingly important tool in medical genetics.[21]

Under the microscope, the chromosomes of human cells appeared as a tangled mess of nearly 50 rod-shaped bodies. During the 1950s, human cytogeneticists developed multiple new techniques that markedly enhanced the visibility of chromosomes. In the laboratory of cytogeneticist T. C. Hsu, the adoption of a hypotonic (low-salt) solution during microscopic analysis helped to spread out the chromosomes so that they were easier to distinguish. This innovation ultimately had an unforeseen impact on the significance of counting in human cytogenetics. Since the 1930s, cytogeneticists had thought the human chromosome number to be 48, though counting often produced a variable result. A new and more concrete conception of the human chromosome number came from the work of cytogeneticists Joe Hin Tijo and Albert Levan, who used Hsu's hypotonic technique and added the chemical colchicine to arrest cells when chromosomes were visible, as well as a "squash technique" that spread chromosomes out in one dimension. In 1956 Tijo and Levan reported that humans possessed 46 chromosomes.[22]

Postwar medical geneticists often pointed to the establishment of the human chromosome number as the origin point of contemporary human cytogenetics. Before 1956 most human cytogeneticists did not consider an accurate count of the chromosomes in a person's cells to be clinically significant. Work done over the next five years, however, identified multiple instances in which chromosome number corresponded with clinical outcome. In 1958 French physician Jérôme Lejeune gave a conference report on a significant cytogenetic finding that linked Mongolism, a common form of mental retardation, to the presence of an extra chromosome. Lejeune first presented his evidence at the Tenth International Congress of Genetics in Montreal, and it was met with a skeptical response. In the following months, Lejeune identified the extra chromosome, later numbered 21, in additional patients clinically diagnosed with Mongolism.[23]

Lejeune first published on the link between Mongolism and an extra chromosome in 1959.[24] That same year, British geneticists Charles Ford and Patricia Jacobs independently reported similar findings, but Lejeune's 1958 conference report had established his priority. Soon thereafter, the identification of a clinically diagnosed Mongolism patient who did not appear to possess an extra copy of chromosome 21 reawakened uncertainty about the reliability of Lejeune's one-to-one link between a microscopically visible chromosomal marker and a distinct clinical disorder. Ultimately, human cytogenetics persevered in defending the clinical value of their findings, demonstrating that an extra copy of chromosome 21 (trisomy 21) did exist in this individual but was hidden from view by a related chromosomal anomaly.[25] Following Lejeune's finding in Mongolism, in the late 1950s and early 1960s, medical geneticists associated multiple disorders with extra or missing chromosomes; included were Turner syndrome, caused by a missing X chromosome in females, and Kleinfelter syndrome, involving an extra X chromosome in males. Disorders caused by extra copies of chromosomes 13 (Patau syndrome) and 18 (Edwards syndrome) were also identified and were found to be lethal during late pregnancy or early infancy.[26]

The one-to-one correspondence between trisomy 21 and Mongolism offered a concrete genetic cause for this well-established clinical disorder, which had first been described almost 100 years earlier. As David Smith put it in 1966, "The greatest enhancement of knowledge in the category of dysmorphology has been the discovery that chromosomal abnormalities represent one mode of pathogenesis." This finding and others like it in the late 1950s and early 1960s pointed to the great potential promise of human cytogenetics in identifying mutations associated with disorders that dysmorphologists believed to have a genetic cause. Lejeune hoped that the identification of trisomy 21 would lead to new treatment options for Mongolism. He dedicated much of his career to curing the disorder, which later came to be known as Down syndrome, but was unsuccessful. Ultimately, the value of trisomy 21 in Down syndrome was limited to diagnosis, which had already been straightforward in clinical patients before 1959. Over the following decades, the most significant impact of trisomy 21 was in the prenatal setting.[27]

In the early 1960s, no method existed for examining the chromosomes of a developing fetus. Amniocentesis could be used to sample fetal cells from the amniotic fluid. However, chromosomes were visible only during cell reproduction, and fetal cells reproduced much too slowly to perform robust cytogenetic analysis. Initially, amniocentesis could be done only late in pregnancy, leaving limited time for diagnosis and a potential preventive abortion. In 1966 physicians Mark W. Steele and W. Roy Breg reported the development of a culture medium that enhanced the reproduction of fetal cells in amniotic fluid, making prenatal cytogenetic analysis possible after a few weeks of culturing. Over the next decade, the uptake of amniocentesis and prenatal cytogenetic diagnosis grew in the United States. Its use was initially limited because of the potential harm that the sampling needle posed to the fetus. Visual ultrasound guidance helped to reduce the chance of injury and allowed the procedure to be more safely performed earlier in pregnancy. As ultrasound became more widespread in the 1970s, so too did amniocentesis.[28]

From the late 1960s, chromosomal analysis following amniocentesis was a significant component of what historian Ilana Löwy has termed the prenatal diagnosis "'dispositif'—a dynamic and constantly evolving array of techniques and approaches which provide information on the fetus during pregnancy." Many of the prenatal tests developed during the 1970s aimed to prenatally diagnose inherited disorders, including Tay-Sachs disease and thalassemia. Often these were not genetic tests but biochemical ones, measuring the amount of certain metabolic chemicals in the fetal urine or blood. Prenatal chromosomal analysis introduced the opportunity to examine the fetal genome directly. Thus, looking for trisomy 21 and other chromosomal abnormalities in the amniotic fluid came to be known as a form of "genetic" testing.[29]

Amniocentesis followed by prenatal chromosomal analysis was a resource-intensive procedure, recommended only for a limited number of pregnancies. Physicians and policymakers targeted women over age 35, who were thought to be at an increased risk for the chromosomal abnormality that caused Down syndrome. A widely circulated narrative about the origins of this age threshold suggested that it represented a balance between the increasing risk of Down syndrome with maternal age and the potential for amniocentesis to harm the fetus. In 2002 genetic counselor Robert Resta demonstrated that the selection of age 35 instead derived from cost-benefit analyses by physicians and policymakers, who sought to balance investments in targeted prevention programs with the expected cost of lifetime care for children born with Down syndrome. Indeed, a risk statistic of the likelihood for amniocentesis to harm the fetus was not established until the late 1970s, by which time age 35 was widely used in obstetric practice.[30]

The age threshold for amniocentesis remained in place in the United States until 2007, when the American College of Obstetricians and Gynecologists recommended that all women be offered access to prenatal genetic testing. Their decision came during a period of significant expansion in the prenatal diagnosis "dispositif," to include new forms of ultrasound and maternal blood testing, which made Down syndrome screening much more broadly available.[31] The policymakers also expressed their increasing confidence that amniocentesis was a very low-risk procedure that should not be withheld in any pregnancy. This decision reflected a larger shift in early-twenty-first-century medical genetics, away from focusing on providing targeted prenatal prevention to specific at-risk populations and toward offering untargeted whole genome screening to all women. Medical geneticists believed that, in some cases, they would identify random disease-causing mutations in fetuses, mutations that could not otherwise have been predicted.

One Mutation–One Disorder

The identification of a cytogenetic marker for Down syndrome was a lauded accomplishment of the "golden age" of human cytogenetics, which stretched from the late 1950s into the early 1960s. Over time, cytogeneticists came up against the limitations of their visual techniques, most notably that the human chromosomes could not be individually differentiated beyond seven groupings. While chromosomal analysis offered a "glimpse" of the entire human genome, it was very difficult for medical geneticists to identify reoccurring mutations among people if these mutations could not be mapped to individual chromosomes and specific locations along them. Human cytogeneticists thus found themselves "in the doldrums" for the rest of the 1960s. Around 1970 new staining techniques gave each human chromosome a distinct banding pattern, aiding cytogeneticists in visually distinguishing them. Cytogeneticists also treated these bands as a set of markers, among which a specific mutation, disease, or gene could be mapped.[32]

With the development of chromosomal banding, efforts to map human genes along the chromosomes expanded rapidly during the early 1970s. Included was an interest in mapping the genetic causes of human diseases. At this time, Victor McKusick was actively laying the groundwork for the growth and spread of medical genetics, in part by calling for a thorough exploration of the human chromosomes. McKusick trained as a cardiologist at Johns Hopkins University, but an early interest in various inherited disorders drew him to medical genetics. Despite a lack of formal training in genetics, in the early 1960s McKusick became a leader in medical genetics and played a significant role in the growing size and prominence of the field. In the 1970s and 1980s, McKusick drew on metaphors to encourage greater public, government, and medical interest in studying human chromosomes; he spoke of cartographic exploration and anatomical study, as well as a need for moon-shot-level investment.[33]

In addition to being a major proponent of human cytogenetic exploration and mapping, McKusick was active throughout the postwar decades in cataloging various human traits, including diseases, that appeared to be associated with a single gene based on their distinctive Mendelian inheritance patterns. His 1966 catalog *Mendelian Inheritance in Man* organized genetic traits by the dominant and recessive modes of inheritance described by botanist Gregor Mendel a century earlier. Dominant traits were expressed when inherited from one parent, whereas recessive traits were activated only when inherited from both parents. McKusick also added X-linked traits, which were expressed clinically when passed down from mother to son. If a disease followed a Mendelian pattern, McKusick believed it to result from a distinct genetic cause.[34]

McKusick was skeptical about the reliability of clinical observations for delineating genetically discrete disorders. Clinical findings, he argued, were often misleading; they led to the inaccurate assumption that disorders were related. In *Mendelian Inheritance in Man,* he wrote, "In medical genetics there is little place for expressions such as 'spectrum of disease,' 'disease A is a mild form, or variant, of disease B,' and so on. They are either the *same* disease, if they are based in the same mutation, or they are *different* diseases. Phenotypic overlap is not necessarily any basis for considering them fundamentally the same or closely related."[35] When medical geneticists sought to delineate disorders, a common mutation was regarded as the most reliable indicator, overruling any apparent overlap in clinical features. In some cases, medical geneticists identified genetic heterogeneity in a disorder that appeared to be a single entity in the clinic. McKusick noted that with genetic heterogeneity, "what at first is thought to be one entity is found to be several clinically similar (i.e., phenotypically similar) but fundamentally (i.e., genotypically) distinct disorders."[36] Even if a disorder looked the same among many clinical patients, medical geneticists considered it to be multiple distinct entities if two or more distinct gene mutations were found to cause the condition in different individuals.

In making these arguments, McKusick pointed back to the bacteriological era a century earlier, when the concept that specific germs caused distinct diseases was established. He viewed genetic medicine in the same framework, arguing, "The principles of genetics force one to think in terms of a *specific* mutation as a *specific* etiologic mechanism resulting in a *specific* disease entity."[37] Aligning the late-nineteenth-century microbiological concept of one germ–one disease with the mid-twentieth-century molecular biology understanding of one gene–one protein, McKusick proposed and defined the ideal model of medical genetics thinking as *one mutation–one disorder*. This was a compellingly simplistic model for thinking about disease, which McKusick understood would not work perfectly in every case. The complexity included instances of what McKusick called genetic "pleiotropism," in which medical geneticists believed that multiple seemingly distinct clinical outcomes were caused by the same mutation. In many instances, including DiGeorge syndrome (described in chapter 5), the identification of a causative mutation led medical geneticists to group variable clinical features as components of the same genetic disease.[38] As a framework for thinking about the cause and prevention of genetic disease, *one mutation–one disorder* served postwar medical genetics well and was influential into the second decade of the 2000s.

The influence of the one mutation–one disorder ideal was reinforced by its underlying circular logic. In order to count as a mutation in medical genetics, a genetic variant needed to cause some discernible pattern of bodily malformation. At the same time, in order to count as a discrete genetic disorder, a condition had to result from a single reoccurring mutation. These were not mutations that increased susceptibility to a disorder, as in the case of cancers. Rather, a mutation in this ideal model of medical genetics caused its associated condition in all individuals who possessed the mutation in sufficient quantity (e.g., one copy for a dominant disorder, two for a recessive one). Uncertainty over what constituted a mutation, or a distinct genetic disorder, animated debates and offered new research directions in medical genetics throughout the postwar period.[39]

The clinical delineation of genetic disorders and attempts to localize causative mutation for these conditions among the human chromosomes constituted a drawn-out process that was beset with unexpected complications and depended on continuous collaboration between physicians and human cytogeneticists. Over time, however, medical geneticists succeeded in establishing links between thousands of genetic mutations and clinical disorders. The parallel development of a chromosomal infrastructure, whereby these associations were located, organized, and later referenced, fundamentally reflected the embodiment of the one mutation–one disorder ideal in medical genetics.

Life Histories of Genetic Disease

Physicians in the postwar period understood diseases as having both a natural and a clinical history. The natural history of a disorder was what happened in the absence of any medical intervention, whereas medical management shaped the clinical history of a disease.[40] These assumptions reflected the ontological perspective of twentieth-century medicine, which treated diseases as real external entities that developed in and affected all individuals in a similar way. Alternatively, late-twentieth-century historians of medicine offered a more constructivist view of human disease, highlighting the role of individuals and societies in making sense of their bodily experiences. As Charles Rosenberg put it, "In some ways disease does not exist until we have agreed that it does, by perceiving, naming, and responding to it." Addressing the complex nature of disease, Rosenberg noted that disease was both a "biological event," which could affect humans and animals alike, and an entity that had been created by society. Thus, "a disease does not exist as a social phenomenon until we agree that it does—until it is named."[41]

From a constructivist perspective, the distinction physicians had drawn between natural and clinical histories was misleading, because no disease category could exist without human intervention. A particular disease, historians of medicine argued, could not be studied independent of human intervention because the disorder's identification and delineation was itself a clinical process, significantly influenced by social presumptions about what counted as a disease. Even at the most basic level, scholars pointed out, drawing distinctions between the normal and the pathological was a fundamentally social process.[42] This book traces the life histories of multiple genetic diseases from the time before their clinical delineation or naming. These disorders were constructed within multiple biomedical research communities, based on particular ways of making visible, measuring, interpreting, and classifying human variation.

The term *life history* is used here in two senses. In postwar medical genetics, disorders had an embodied life history within affected individuals, composed of various developmental features from the prenatal period to death. Medical geneticists worked to generalize about the embodied life history of a disorder, but they knew that the pattern of malformation would appear somewhat differently in each patient. In addition, medical geneticists understood disorders as having a conceptual life history, with no clear beginning or end. This life history was made up of evolving medical and social understandings of a condition's defining characteristics and cause, as well as how it might be diagnosed, treated, and prevented. Historically, medical professionals played the most prominent role in characterizing the life histories of disease. However, in the late twentieth century, patients and their advocates began to play a greater role in shaping perceptions and understandings of the life history of their conditions.[43]

While I have distinguished here between the embodied and the conceptual life histories of genetic disease, in practice I believe it is futile to draw rigid boundaries between them. Postwar understandings of the conceptual and embodied life histories of a disorder continuously relied upon and informed one another. For physicians, the interpretation of a disorder's embodied life history was directly influenced by contemporary assumptions about its conceptual life history, and in turn observations of the embodied life history of a disorder fed back into conceptual understandings of it. Mindful of this, I use the term *life history* to refer to the embodied and conceptual aspects of a disorder without specification. Blurring the lines between the embodied and the conceptual serves as a helpful reminder that disorders exist as part of a messy juxtaposition between bodies, ideas, and social structures.

Late-twentieth-century debates over the adoption and use of medical terminology were also central to shaping the life histories of genetic disorders. I use the terms *disease, disorder,* and *condition* interchangeably. However, in different circles, these words have been linked to more specific connotations. For instance, some have interpreted the term *disease* to imply a well-established, clinically serious ailment of known cause. As McKusick noted in *Mendelian Inheritance in Man,* "There is little rhyme or reason to the use of *disorder, disease, syndrome,* and *anomaly. Disease* often has unhappy connotations to the layman; *disorder* or *syndrome* is more satisfactory. Some suggest that a disorder be called a disease when the basic defect becomes known."[44] McKusick knew well, after decades of cataloging disorders, that in postwar medical genetics there was no ultimate authority deciding what a disorder should be named. Medical naming evolved like any other language, through use, misuse, and disuse. One could make general suggestions (and McKusick made many), but it was hopeless to try to standardize when physicians should use the designation *disease* or *disorder* instead of *syndrome.*

The primary disorders examined here are named as syndromes. The syndrome concept was an important one in postwar medical genetics, even though not every genetic disease that had syndromic characteristics was explicitly named as one. Medical geneticists understood syndromes as having a single cause that affected multiple body systems. Crucial to the syndrome concept was a tolerance for variability. Syndromes were characterized by a hierarchical list of features, not all of which (including even the most common and distinctive) needed to be present for diagnosis. Identifying a syndrome required attention to various subtle details and also significant flexibility in one's preconceptions of what a syndrome looked like.[45]

McKusick's dissatisfaction with delineation through similarity in bodily features derived from his familiarity with the complex clinical process of syndrome diagnosis. Nonetheless, the characterization of a new syndrome in medical genetics most often started with the identification of a clinical population with notable features in common.[46] Delineation gained momentum when other physicians reported the same condition in other patients. Medical geneticists also sought a causative mutation among affected individuals. The identification of a common mutation was a significant moment in the construction of a disease's life history. Medical geneticists saw it as evidence that a disorder was distinct, diagnosable, and preventable. Postwar scholars and social advocates criticized the tendency of medical geneticists to reduce disorders, and those affected by them, to genetic mutations.

Genetic Reductionism and Disability

In the late twentieth century, many scientists referred to the human genome as "the book of life" and as a "blueprint" for health and development. These metaphors conveyed the significance that researchers and their financial supporters attached to human gene mapping and genomic sequencing. As sociologist Dorothy Nelkin and historian Susan Lindee have documented, the gene-for concept was very powerful in postwar society, whether it provided an explanation for cancer, fashion sense, sexuality, or intellect. Even as results from the Human Genome Project in the late 1990s led scientists to acknowledge that many human traits were about more than genotype, reductive and deterministic logic retained its simplifying appeal in public discourse.[47]

Postwar medical genetics was based on the premise that variations in the genetic makeup of individuals could reveal something about their present or future health. Based on this, medical geneticists promised that improvements in genetic knowledge and testing would enhance public well-being. From an alternative perspective, critics of genetic medicine were concerned about the tendency of its practitioners to reduce the health outcomes of individuals in otherwise complex cultures and variable environments to single genetic variants. To situate genes as one of the most significant factors in shaping public health, they suggested, was to undercut the significance of disparities in wealth, environment, and access to health care.

In the early 1990s, epidemiologist Abby Lippman, in her article "Led (Astray) by Genetic Maps," criticized efforts to locate the causes of human disease in the genome. Medical geneticists, she lamented, defined illness as a problem of individual genes, rather than a condition shaped by society.[48] Many other scholars and social critics of genetic medicine similarly held that what counted as normal or healthy was not written into our genes but was instead a social decision. They demonstrated that value-based judgments were what determined which conditions were acceptable and worthy of treatment, rather than targets for genetic prevention or eradication. The identification of more genetic variants, some scholars argued, only meant that there would be more disease entities, many of which would be characterized as abnormal. Instead of providing more "choice," the existence of additional diagnosable disorders was seen as likely to increase parental anxiety and the complexity of decisions in the prenatal context. More broadly, critics argued, the expansion of testing options had opened wider the "Backdoor to Eugenics." While reproductive decisions were left to parents rather than governments, scholars noted that perceptions of what was abnormal or unacceptable were still shaped by social norms.[49]

During the 1990s, debates took place within the medical profession and within the increasingly influential disability rights community about which conditions were serious enough to justify preventive abortion. The traits that could be tested for prenatally at this time ranged from extra fingers to disorders that were often lethal in early childhood.[50] Discussions about where to draw a line of acceptability among these conditions were rarely fruitful, because of a diversity of opinions and perspectives on disability. Many in the disability rights community argued that changing social perceptions and support structures could transform traits that many considered disabling into "neutral" traits. Rather than accepting a medical perspective of disability as a form of abnormality or individual deficit, requiring treatment or prevention, the disability rights community adopted a social model, framing disability as a condition created by society's lack of accommodation and acceptance of particular physical and personality traits. Efforts to fully integrate disabled persons into society, activists suggested, would eliminate the need and demand for prenatal testing.[51]

For decades, many physicians had taken for granted the idea that preventing disorders like Down syndrome was in the best interest of affected individuals, families, and society. By the 1990s, however, families affected by disabling genetic disorders increasingly decided to forgo available prenatal testing that would allow them to avoid having additional children with the same condition. Having gotten to know their child as an individual, some no longer regarded the child's condition as a disability or something to be prevented. Many families also viewed the pursuit of prevention by other parents as both an insult and a threat to their own child, as well as to the disability community. Some saw preventive abortion for their child's condition as an expression of discrimination and suggested that the availability and acceptability of prenatal diagnosis reduced whole people down to single genetic traits, overshadowing everything else that the individual had to live for and offer. In addition, families worried that if most parents chose abortion when faced with their child's disability, this would in time obliterate their community and undercut social as well as government support systems for the condition.[52]

Late-twentieth-century disability scholars and bioethicists argued that every child faced challenges in her life and that parents who believed that they could use prenatal testing to achieve perfection through the selection of traits fell victim to a profoundly misleading fallacy. No genetic test could guarantee a normal child, because normality and abnormality, like disease categories, were socially constructed and fluid classifications. As bioethicist Evelyne Shuster put it in 2007, as the scope of prenatal genetic testing increased, the greatest risk that potential parents faced in their search for a normal or perfect child was "no baby at all."[53] Indeed, by 2010, with the increasingly dense and still growing infrastructure for genetic diagnosis being applied in its entirety in the prenatal setting, parents were presented with an ever higher bar for what constituted a "wellborn" child. As prenatal genetic diagnosis became less targeted and was more broadly applied for uncovering any potential abnormality, parents who had sufficient means and chose to pursue testing were faced with a preventive regime in which it had become increasingly difficult to receive medical reassurance that theirs would be a healthy baby.

Overview of Chapters

The following six chapters trace, roughly chronologically, the development of medical genetics and its chromosomal infrastructure for diagnosis and prevention between 1970 and the 2010s. Odd-numbered chapters offer in-depth historical accounts of the life histories of representative genetic disorders. Each disorder was chosen to reflect one or more aspects of the evolution of postwar medical genetics in the laboratory and the clinic. Even-numbered chapters focus more specifically on the development of new techniques in medical genetics. Novel approaches to making genetic diseases visible played a significant role in the construction of their life histories.

Chapter 1 examines an important organizational period in medical genetics around 1970 and focuses on the practices and collaborative efforts of dysmorphologists and human cytogeneticists. Both of these specialties contributed to the delineation of an inherited form of intellectual disability later called fragile X syndrome. Dysmorphologists identified fragile X syndrome as a distinct pattern of malformation in institutionalized populations, while human cytogeneticist Herbert Lubs demonstrated that the disorder was associated with a particular chromosomal abnormality. Medical geneticists found the bodily and chromosomal expression of fragile X syndrome to be surprisingly variable. During the 1970s and 1980s, researchers faced many complications as they attempted to fit the disorder within the one mutation–one disorder ideal of medical genetics and tried to make fragile X syndrome amenable to reliable targeted prevention.

In 1969, when Herbert Lubs first described the cytogenetic marker later associated with fragile X syndrome, he had no direct means of visually identifying on which chromosome it was located. As chapter 2 explains, over the following years, new staining techniques made visible a banding pattern along the human chromosomes, which cytogeneticists used to distinguish them. Human cytogeneticists also developed a standardized system for naming these bands. This was the basis of the chromosomal infrastructure that became fundamental to human genome mapping. During the 1970s, medical geneticists worked to identify the location of various genes and disease-causing mutations along the chromosomes. Foremost among them, Victor McKusick positioned the human gene and disease map, built on the chromosomal infrastructure, as the central work object of medical genetics. In doing so, he helped to create a sense of common interests and goals among a heterogeneous set of researchers, funders, and public interests.

As part of his ongoing promotion of human genome mapping during the early 1980s, McKusick spoke of disease-causing mutations as comprising the human genome's "morbid anatomy." Chapter 3 looks at how McKusick and other influential medical geneticists established the localization of clinical disorders within the human genome's morbid anatomy as central to the work of medical geneticists. In particular, McKusick positioned Prader-Willi syndrome as an exemplar of the ability in medical genetics to identify the genomic location of disorders using new high-resolution cytogenetic techniques. As in fragile X syndrome, however, variable chromosomal findings complicated the one mutation–one disorder ideal and called into question the reliability of presymptomatic diagnosis. In response, medical geneticists began to look to molecular approaches for assistance in resolving this issue.

Chapter 4 examines the integration of practitioners and techniques from molecular biology into human cytogenetics. Faced with the complexity and uncertainty of visual chromosomal analysis, medical geneticists turned to molecular techniques to provide a more "objective" approach to disease diagnosis. Human cytogeneticists believed that even when a mutation was too small to be made directly visible under the microscope, DNA hybridization techniques from molecular biology could be used to determine the presence or absence of a chromosomal mutation. During the 1980s and 1990s, human cytogenetics was molecularized, but chromosomal analysis continued to have a central role in gene and disease mapping. Human chromosomes remained the primary organizational units of medical genetics and the basis of its diagnostic infrastructure, because they were familiar and legible to so many of its practitioners. The integration of molecular approaches into human cytogenetics ultimately led to the development of molecular cytogenetics, a new hybrid specialty in medical genetics that sought to bridge the gap between visible chromosomal markers and molecular genes.

Throughout the 1980s, McKusick noted that new techniques in human cytogenetics had greatly enhanced the ability of medical geneticists to isolate genetically discrete disorders "in pure culture." Just as Koch's postulates had facilitated the isolation of specific germ causes of disease during the late nineteenth century, medical geneticists in the late twentieth century began to study discrete genetic diseases by identifying their causative mutations. McKusick argued that genetic markers were the most reliable evidence for medical geneticists as they worked to "lump" or "split" disorders that looked similar in the clinic. Chapter 5 describes how the life histories of two genetic diseases, DiGeorge syndrome and velo-cardio-facial syndrome, were merged into one based on the identification of a common mutation. While medical geneticists eventually agreed that these historically distinct disorders were one, the separate names and institutions that had been built around them did not so easily collapse into a singular identity. The convergent life histories of these syndromes demonstrated that disease identities and communities were built around much more than mutations, even as medical genetics sought to reduce disorders to mutations.

Chapter 6 deals with the expansion in density and scope of genetic testing after the completion of the Human Genome Project in 2000. At this time, prenatal diagnostic approaches were targeted to preventing disorders in individuals or families that were at known risk. Medical geneticists observed, however, that most disease-causing mutations occurred randomly and were unlikely to be tested for prenatally within this framework. The disorders that random mutations caused were individually rare but relatively more common when considered as a group in the whole population. Mindful of this fact, in the 1990s molecular cytogeneticists turned their attention to developing techniques, including DNA microarray, which could be used to test for hundreds or thousands of random mutations simultaneously. Microarray screening, combined with knowledge from the chromosomal infrastructure, allowed for much broader testing regimes, promising parents "the whole picture" but often providing only increased uncertainty.

Reflecting on the impacts of new molecular technologies and the Human Genome Project, some scholars have argued that a fundamental transition took place in biomedicine after 1980, shifting the focus to primarily molecular conceptions of disease. Resisting this interpretation, I examine in the epilogue the continued importance of clinical and chromosomal styles of thought and representation in twenty-first-century medical genetics. The era between 1970 and the present involved not a transition from one style of thought to another, but rather the ongoing development of a genomic gaze, incorporating clinical, chromosomal, and molecular understandings and depictions of disease.

Chapter 1

Genetics Detectives

During the mid- to late twentieth century, medical geneticists divided into hundreds of distinct disease categories individuals who, 50 years earlier, would have been called feebleminded. Many of these novel disease categories were delineated and constructed by medical geneticists as part of their examinations of mental deficiency in institutional settings. British physician Lionel Penrose's Colchester Survey, conducted during the 1930s, was among the first and most influential of these studies. Penrose examined 1,280 residents of the Royal Eastern Counties Institution, as well as thousands more of their family members, seeking to explore the relationship between genetic and environmental factors in causing mental deficiency. Penrose's 1938 report on the Colchester Survey strongly refuted the more simplistic assumptions of early-twentieth-century physicians and eugenicists about the cause and inheritance of mental deficiency.[1]

Despite its more nuanced perspectives on mental deficiency, Penrose's report offered little in the way of more effective modes of prevention. During the decades that followed the Colchester Survey, medical geneticists acknowledged that they possessed few tools for preventing mental deficiency. Most believed that traditional eugenics approaches, such as discouraging reproduction and forcible sterilization, could help. However, these blunt measures were inefficient and were increasingly falling out of favor in the United States and the United Kingdom. And anyway, as Penrose and his colleagues knew well from clinical experience, mental deficiency occurred in many children whose parents appeared normal and thus were not prevented by such programs from reproducing.[2]

Biologist J. B. S. Haldane, in his preface to Penrose's 1949 book *The Biology of Mental Defect*, compared contemporary efforts to prevent children from being born with mental deficiency to Leonardo da Vinci's fifteenth-century dream of building a flying machine. The means of controlling mental deficiency were not yet apparent, but Haldane professed optimism that the study of human genetics would eventually provide valuable tools.[3] At this time, patients with mental deficiency were categorized primarily by the severity of their condition. In the United States, the term *idiot* referred to those with the lowest IQ, followed by *imbecile* for the midrange and *moron* to describe patients with relatively minor impairment.[4] Postwar medical geneticists believed that these categories were too broad to correlate with any discrete genetic disorder. Beginning in the 1960s, they instead turned their attention to identifying distinct conditions in which mental deficiency was part of a reoccurring pattern of malformation, suggesting a single genetic cause.

Physicians and geneticists gained confidence in the promise of genetic testing for the prevention of mental deficiency during the "golden age" of human cytogenetics, in the early 1960s. Most notably, the chromosomal link between trisomy 21 and Down syndrome was seen as an important model for future research. Over subsequent decades, physicians and geneticists worked collaboratively to link discrete phenotypic patterns from the clinic to chromosomal mutations that were visible in the laboratory.[5] In this process, newly developed subspecialties of dysmorphology and human cytogenetics rose to prominence. Many investigators in these areas were influenced by the medical genetics ideal of one mutation–one disorder. Medical genetics researchers occasionally struggled to make disorders fit with this model. After years of work, many of these conditions were explained in terms of a one-to-one association, providing medical geneticists with the necessary evidence and reliability to pursue prenatal diagnosis and targeted prevention. Fragile X syndrome, the focus in this chapter, offers a representative example.

The chromosomal mutation in fragile X syndrome appeared unique to medical geneticists in important and complicated ways. Most disease-causing chromosomal abnormalities occurred randomly during reproduction and were not passed down in families. Fragile X syndrome was inherited. Medical geneticists generally assumed that inherited disorders were caused by single-gene mutations, which could not be seen under the microscope. Fragile X syndrome was both inherited *and* involved a visible chromosomal mutation. For medical geneticists, this bridged a significant gap between prenatally diagnosable chromosomal disorders and inherited disorders. Ultimately, to make the disorder amenable to prenatal diagnosis and targeted prevention, medical geneticists still had to resolve the unusual inheritance pattern and behavior of the fragile X mutation. This multidecade process reflected broader efforts in postwar medical genetics to identify disorders that could be made reliably visible, and thus preventable, prenatally.

Delineating fragile X syndrome during the 1970s and 1980s involved a continuous interchange of findings between dysmorphologists and human cytogeneticists. The identification of a pattern of bodily malformation went hand-in-hand with the laboratory recognition that its associated chromosomal marker was in fact a mutation. Collaborative synergy between these clinical and cytogenetic ways of seeing was enhanced by a growing network of common resources, which organized and characterized patterns of malformation within the one mutation–one disorder ideal of medical genetics. Along with practitioners themselves, reference texts in human cytogenetics and dysmorphology increasingly referred to one another. The hybrid work of these postwar medical genetics subspecialties aimed to construct the life histories of many genetic diseases, with the hope of facilitating targeted prenatal prevention.

A Marker for Mental Defect

In 1967 Yale–New Haven Hospital admitted a one-year-old boy to investigate the causes of his developmental and cognitive delay. At Yale, physician and geneticist Herbert Lubs was engaged in a large study, in which the chromosomes of all newborns at the hospital were being scanned for visible abnormalities. Much research in human cytogenetics at this time was driven by a desire to better understand what abnormality among the chromosomes looked like. How, cytogeneticists wondered, could pathological mutations be reliably distinguished from normal variation? They believed that ultimately such distinctions could be appreciated only in the context of large studies of a variety of individuals, both sick and healthy.[6] Lubs had been interested for years in the chromosomal basis of human disease and believed that the one-year-old's condition might represent a genetic disorder. He entered the child into the larger Yale–New Haven study as "YNHH 65-91-42" to have his chromosomes studied for markers of abnormality.[7]

Lubs's chromosomal analysis identified a visible variant, which he characterized as "an unusual secondary constriction . . . [giving] the appearance of large satellites." Under the microscope, the chromosome appeared to have two appendages attached to it, making it quite distinct from the boy's other chromosomes.[8] Having identified an unusual-looking chromosome, Lubs's next step was to analyze the boy's parents, to determine whether this visible genetic variant was limited to him or was being passed down through the family. Such studies, which began with a clinically affected individual and then traced genetic and bodily traits through their family pedigree, were central to the thinking and practices of postwar medical genetics and strongly shaped conceptions about the genetic basis of disorders involving mental deficiency.[9]

Examination of the boy's immediate family revealed the same unusual genetic marker in his intellectually normal mother. Usually such a finding would suggest that the variant was not responsible for the clinical disabilities seen in the child, since it did not appear to affect his mother. However, this boy had an older brother who showed cognitive and developmental delay and who also possessed the same visible aberration. It remained unclear whether this chromosomal marker was a form of benign cytogenetic variation or a disease-causing mutation. To help answer this question, Lubs conducted a broader survey of the family. His analysis found two male relatives who were also mentally disabled and possessed the same chromosomal marker. No female members of the extended family showed any mental deficiency, but an aunt of the two boys was also seen to possess the same secondary construction. The finding that this genetic trait was correlated with mental deficiency only when seen in males suggested to Lubs that it was inherited in the X-linked recessive Mendelian pattern.[10]

Lubs and his colleagues in human cytogenetics understood Mendelian traits to be discrete genetic entities that were passed down through a family on chromosomes. One copy of each chromosome was inherited from a child's father and one from the mother. Both parents had two copies, and one from each was passed on to offspring at random. Human cytogeneticists believed that sex chromosomes acted differently. Females had two copies of the X chromosome, and males had one. As a result, genetic defects inherited on the X chromosome in males were expected to be clinically expressed, while in females expression was often prevented or mitigated by the presence of a second, normal, copy of the X chromosome. In the late 1960s, human cytogeneticists were unable to visually distinguish the X from chromosomes 6–12, which were of similar size and shape.[11] Lubs relied on a series of additional laboratory examinations to demonstrate that the unusual secondary constriction did indeed occur on the X chromosome.[12]

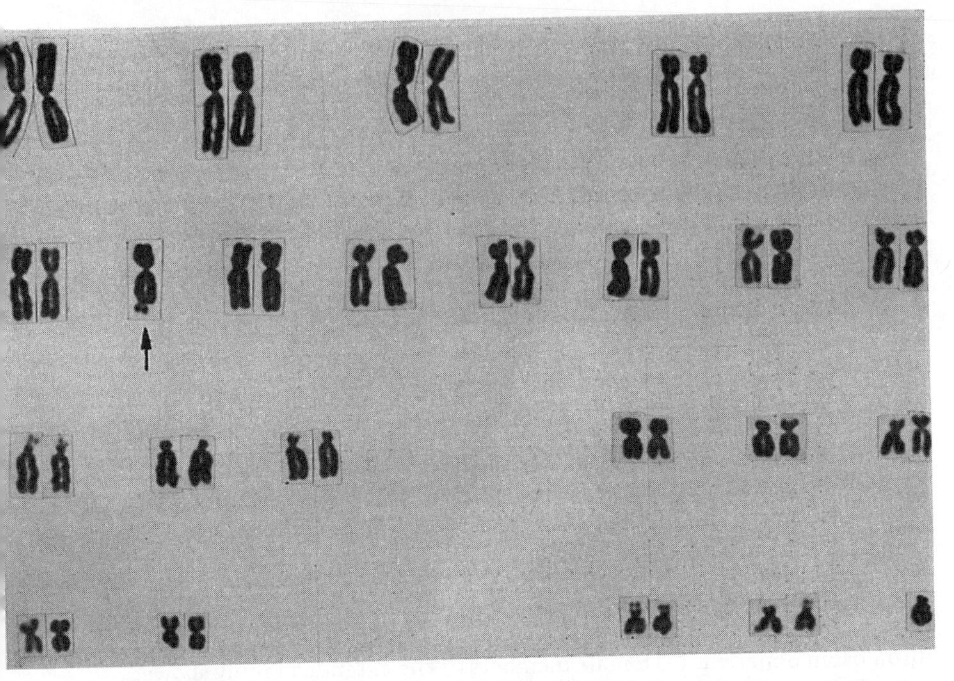

Karyotype highlighting the secondary constriction on the marker X chromosome. *Source:* H. A. Lubs, "A Marker X Chromosome," *American Journal of Human Genetics* 21, no. 3 (1969): 235. Copyright Elsevier.

Normally, Lubs knew, mutations in Mendelian traits were not visible, because they involved DNA-level changes in individual genes, which could not be seen in microscopic observation. During the 1960s medical geneticists recognized two primary categories of disease-causing mutations: those involving chromosomes, and those affecting individual genes. Penrose characterized this distinction by referring to "mistakes of an imaginary printer"—affecting the DNA code—and "mistakes of a binder," involving chromosomes. Similarly, influential medical geneticist Victor McKusick distinguished between syndromes caused by large chromosomal aberrations and those caused by Mendelian inherited traits.[13] The X chromosome variant that Lubs had identified, like others described in subsequent chapters, fell between these two categories. Visible abnormalities like this one suggested to medical geneticists that the chromosomal location and cause of single-gene disorders, which were responsible for the ma-

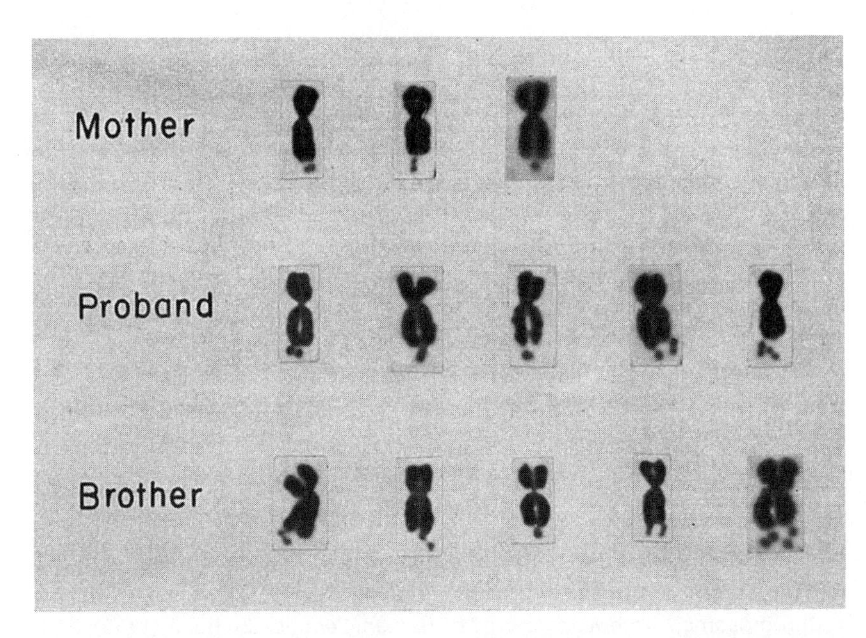

Marker X chromosomes from the original patient (proband), his similarly affected brother, and their intellectually normal mother. *Source:* H. A. Lubs, "A Marker X Chromosome," *American Journal of Human Genetics* 21, no. 3 (1969): 236. Copyright Elsevier.

jority of inborn genetic diseases, might in some cases be discernible under the microscope.

Lubs published his findings in the May 1969 issue of the *American Journal of Human Genetics*. Aware of the potential significance of his discovery, he began by proclaiming, "Descriptive human cytogenetics is entering a new and important phase."[14] The secondary construction, or "marker X chromosome," as he called it, was a much more subtle variant than those, including trisomy 21 in Down syndrome, that had previously been associated with mental deficiency. Even more importantly, this variant was inherited. This had not been the case with larger chromosomal abnormalities associated with other disorders. Therefore, Lubs stated, "The most immediate medical consequence of these studies has been the opportunity to offer amniocentesis and therapeutic abortion to two women in this family if the marker chromosome is found in a male fetus." Indeed, while it was unclear how this chromosomal abnormality caused the disorder, it

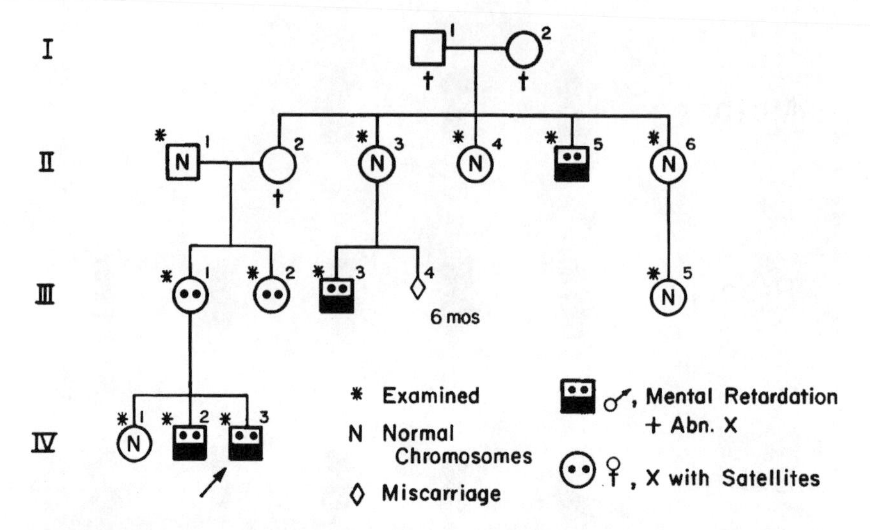

Pedigree of the extended family, showing mentally retarded males with the marker X chromosome in multiple generations. *Source:* H. A. Lubs, "A Marker X Chromosome," *American Journal of Human Genetics* 21, no. 3 (1969): 232. Copyright Elsevier.

seemed to provide a visible marker that could be used to facilitate the targeted prevention.[15]

The clinical utility of amniocentesis for prenatal cytogenetic diagnosis had been demonstrated only recently, in 1966, and very few clinical disorders could be identified chromosomally at this time.[16] Unlike Down syndrome—which was the primary target of early prenatal diagnosis—the "marker X" form of mental deficiency was expected to reoccur in 50% of male births from women who already had an affected child. This made the disorder an ideal target for prenatal testing. In a follow-up report on this initial family, published 15 years later, Lubs found that the mother of these mentally deficient brothers did later have a third son who was normal. There was no indication that she pursued the opportunity for prenatal diagnosis.[17]

Despite the promises that the marker X chromosome offered, many years passed before other human cytogeneticists saw the chromosomal abnormality that Lubs reported in 1969. The marker X chromosome did not behave in the ways that Lubs and other clinical researchers had hoped. Human cytogeneticists found that this and other similar chromosomal abnormalities had to be properly induced in the laboratory in order to be made microscopically visible and clinically interpretable. During the 1970s, cytogeneticists developed new approaches for making such chromosomal abnormalities reliably visible. In some cases chemical manipulation was used to coax chromosomal markers into view and thereby make the diagnosis of a particular disorder possible. The life histories of many genetic diseases were drastically reshaped in the late twentieth century by these new ways of seeing and making sense of chromosomal abnormalities. During the 1970s, the visual approaches of different subspecialties in medical genetics were aligned with the goal of improving diagnosis.

Organizing Medical Genetics

The years around 1970 were important organizationally for genetic medicine. Leaders in the field published several new reference textbooks, which grew significantly in size and influence with revised editions. These texts offered standardized approaches for recognizing and categorizing bodily patterns of malformation and their links to genetic mutation. The goal of each was to expand the reach of genetic medicine to a wider range of clinicians and geneticists. The great majority of clinicians at the time thought of genetic disorders (aside from Down syndrome) as rare birds, most of which they were unlikely ever to see. However, the authors of these texts claimed that genetic diseases, while individually rare, were a significant cause of human disease as a group, and thus a topic with which *all* physicians needed to be familiar.[18]

The most significant organizational text for genetic medicine during the 1970s was Victor McKusick's *Mendelian Inheritance in Man* (*MIM*). The first edition of *MIM* was published in 1966, and the catalog went through three new and significantly expanded editions in 1968, 1971, and 1975. McKusick organized human genetic traits by their mode of Mendelian inheritance: dominant, recessive, or X-linked. For cataloged disorders, *MIM* described associated bodily features and provided a diagnostic history with relevant citations. McKusick assumed that each of the disorders he cataloged in *MIM* had a discrete genetic cause. Thus he promoted the one mutation–one disorder ideal of medical genetics. McKusick's catalog grew quickly during the 1970s, particularly as new techniques based on chromosomal manipulation and analysis helped to identify the locus of many human genes and diseases.[19]

In *MIM,* disorders were grouped by Mendelian inheritance pattern and alphabetized within the groups. This meant that a physician could look up a disorder in *MIM* only if he already had a name. *MIM* was a useful catalog of known disorders but was of limited value in the diagnostic process. To help fill this gap, pediatrician David W. Smith published *Recognizable Patterns of Human Malformation* in 1970, as a clinical guide to diagnosing disorders based on bodily patterns of malformation. A few years earlier, Smith had coined the term *dysmorphology* for the study of abnormal tissue development. Central to Smith's approach to diagnosis was a systematic examination of the body, with a focus on singular causes of malformation. Like McKusick, Smith assumed that these causes were primarily genetic.[20]

In *Recognizable Patterns of Human Malformation,* Smith organized known malformation syndromes by their primary physical defects, including abnormalities in bodily growth, muscle development, facial features, fingers, and toes. An additional organizing category that Smith included was abnormalities in chromosome number, under which Down syndrome (trisomy 21) was listed. In chromosomal disorders, abnormalities that were visible under the microscope were considered to be the primary genetic cause from which a distinct pattern of malformation arose. For each malformation syndrome, Smith described major and minor bodily malformations and, most importantly, provided pictures of affected individuals. His textbook was meant to increase clinical familiarity with and recognition of disorders and their various subtle traits, while also encouraging physicians to think about the singular causes—mostly genetic—of these patterns.[21]

At this time, dentist Robert Gorlin also produced multiple reference texts aimed at helping physicians to identify patterns of bodily malformation. They included *Syndromes of the Head and Neck* (1964) and a 1970 text titled *The Face in Genetic Disorders.*[22] These reference texts focused on abnormalities of the head, neck, and face as significant indicators of particular developmental patterns of malformation. Like Smith's catalog, *The Face in Genetic Disorders* was primarily aimed at helping clinicians make a diagnosis based on recognizing patterns of bodily malformation, and again it expected a genetic cause. This volume also included one-page descriptions of many genetic disorders, with images, but they were organized by mode of inheritance, as in *MIM,* rather than by categories of bodily malformation.

Gorlin also listed, separately from Mendelian disorders, syndromes associated with chromosomal abnormalities.[23] Each of these texts contributed to what historian Susan Lindee called the "cataloguing imperative" of postwar genetic medicine.[24] Used together, they aided clinicians in identifying patterns of malformation and associated genetic disorders that were rarely seen in the clinic.

Medical geneticists have often compared dysmorphology to detective work. Jon Aase, a colleague of Smith's, described the skill set of dymorphologists as being similar to Sherlock Holmes's. As he put it in 1981, each arrives on the scene after the main causative event (for Holmes a crime) has taken place and attempts to identify and piece together clues into a coherent story. In this process, suggested Aase, "perhaps the most striking and immediately useful [technique] is the ability to observe small variations. . . . Inconsequential in themselves, these variants, usually in combination, can point to more significant and far-reaching developmental aberrations."[25] The reference texts of this era offered a systematic way of collecting and analyzing these clues, toward making a genetic diagnosis. They contributed to the delineation of many new disorders in the 1970s and 1980s, including fragile X syndrome.

Recognizing Patterns of Mental Deficiency in the Clinic

State institutions for the mentally deficient were important sites of knowledge production for biology and medicine during the nineteenth and twentieth centuries.[26] Beginning in the mid-nineteenth century, institutions in France, England, and the United States began to isolate mentally deficient individuals from the larger psychiatric population. Asylum laws in England at this time required that all residents undergo extensive medical assessment, including many anthropometric measurements. John Langdon Down, a young physician and medical director of the Earlswood Asylum during the 1860s, was kept very busy compiling clinical records of his more than 400 patients. Historian David Wright has argued that this continuous analysis of the physical traits influenced Down's thinking about distinct forms of mental impairment. In 1866 he reported on a distinct grouping of mentally deficient patients that he called Mongolian idiots, based on their facial features, which Down believed resembled those of the Mongolian race. Similar individuals were described in other institutions and diagnosed with Mongolism over the following decades.[27]

Many clinicians followed in Down's footsteps. Mongolism (later renamed Down syndrome) was a major focus of Penrose's survey of patients at the Royal Eastern Counties Institution during the 1930s. In the late 1960s, Gillian Turner, a pediatrician in the Australian state of New South Wales, was similarly involved in assessing patients who were institutionalized because of mental retardation. Like other clinicians at this time who studied the genetic basis of disease, Turner became interested in delineating particular disorders among large populations of mentally retarded individuals. Influenced by Smith's and Gorlin's early work on recognizable patterns of bodily malformation, Turner developed a systematic approach for assessing her patients, which involved collecting information about family history, metabolic disorders, heights, weights, and head circumferences, among other data points. All of this information was stored on computer punch cards for later analysis. In the course of her work, Turner became familiar with an extended family in which many individuals, but only males, were affected by mental retardation. Eventually, colleagues introduced her to four additional families in which only males were affected, a pattern suggesting that an X-linked genetic trait might be involved.[28]

Interested in identifying patterns of bodily malformation, Turner began searching the medical records and images of these institutionalized patients who seemed to be affected by X-linked mental retardation for other notable features that they had in common. What she found surprised her: "It suddenly dawned on me that they all had just nothing exceptional, and that was the exceptional feature—they were all physically normal."[29] Indeed, among the cases of apparently X-linked mental retardation that Turner surveyed, many patients did not show any other remarkable clinical features. Individuals who were mentally retarded but otherwise normal-looking were thus set apart from those who had disorders like Down syndrome, which involved distinct facial features, and phenylketonuria (PKU), a metabolic disorder. Turner hypothesized from her patients that X-linked forms of mental retardation were more likely to involve few distinguishing clinical features. Returning to her punch card catalog of 147 institutionalized patients, she identified a group of 17 physically normal males. Among them, 60% had a mentally retarded brother or other male relative, suggesting X-linkage.[30]

Turner published her findings in a letter to *Lancet* in 1970 and suggested the name "Renpenning syndrome" for boys who had this physically

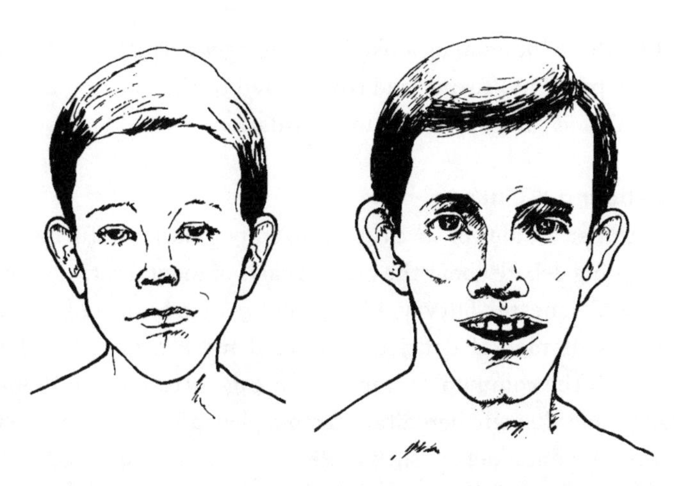

Composite drawings of a child and a young adult with fragile X syndrome. *Source:* J. C. Mixon and V. G. Dev, "Understanding the Fragile X Syndrome," *Alabama Journal of Medical Sciences* 21, no. 3 (1984): 284. Reprinted with permission from V. G. Dev.

"nonspecific" form of X-linked mental retardation. She was acknowledging earlier work by Hans Renpenning, who in 1962 reported on a family in Canada with X-linked mental retardation but no other major distinguishing physical features. Renpenning, a medical student at the time of his report, had previously worked with physician Henry Dunn studying a family affected by apparently X-linked mental retardation in British Columbia. When Renpenning returned home to Saskatchewan, his father pointed him to another similarly affected local family, which became the basis of his 1962 report. This article came out a year before the results of Dunn's study, leading to Turner's assumption that Renpenning deserved priority in naming the syndrome.[31]

In ensuing years, Turner, seeking to widen her survey, began visiting other institutions for mentally retarded males throughout New South Wales, looking for affected brother pairs. Like Down, Penrose, and many other physicians before her, Turner benefited in her research from the concentration of mentally retarded individuals in state-run institutions. In other locations, she once again found that the majority of mentally retarded brother pairs were otherwise normal in appearance. Turner's findings pointed to two patterns among these individuals. One was genetic: the cause of this disorder appeared to be inherited on the X chromosome. The other was bodily: these individuals looked surprisingly normal.[32] In further defining these patterns, Turner had contributed to the ongoing delineation of a new X-linked category of mental retardation.

Debating X-Linkage

With her studies of institutionalized populations, Turner entered into an ongoing debate about the genetic basis of mental retardation. Penrose, in his Colchester Survey, like others before him, had found that severe forms of mental deficiency affected more males than females. Breaking with the common conceptions of eugenicists in the 1930s, Penrose argued that discrete hereditary factors played a dominant role in just a few forms of mental defect, such as PKU.[33] In his revised 1963 edition of *The Biology of Mental Defect,* Penrose also continued to resist an X-linked genetic explanation for the widely documented male excess of mental retardation, suggesting that "in general, the genes on the X chromosome do not play any greater part in the causation of mental defect than might be supposed from the fact that there are 22 autosomes to one sex chromosome in man."[34] Rather, Penrose suggested that X-linkage accounted for only a small portion of the total incidence of mental deficiency. And indeed, in the early 1960s, only a few clear instances of X-linked mental retardation had been reported in the medical literature. Within a decade however, Turner and other clinicians began to identify new examples.

A major contributor to studies on the prevalence of X-linked mental retardation was Robert Lehrke, who, after a career as a school psychologist, became a PhD student at the University of Wisconsin, Madison, during the 1960s. Lehrke had an experience similar to Turner's with institutionalized populations of mentally retarded children, and he had also noticed a prevalence of similarly affected brother pairs among them. Mindful of these related inheritance and clinical patterns, he too became interested in exploring the genetic basis of mental defect. Based on a study of five families from Wisconsin and Minnesota, Lehrke argued in his doctoral dissertation that 25–50% of all mental retardation was caused by X-linked genes.[35]

Lehrke's interpretations proved to be controversial. They caused a rift between two of his dissertation examiners, geneticist James Crow and psychologist Rick Heber, which led to the need for revisions that undercut his argument. Later, in 1971, when Lehrke's results were published in the

American Journal of Mental Deficiency, his arguments once again drew resistance. Immediately following Lehrke's article, the editors published multiple responses from prominent researchers in the field, questioning the representativeness of his data set and the accuracy of his interpretations. Penrose was a very influential figure in mid-twentieth-century medical genetics, and his opposition to X-linked explanations for mental retardation held significant sway in the field.[36]

In making his argument, Lehrke had drawn on his experiences as a school psychologist, as well as the few already existing studies of X-linked mental retardation. These included a 1943 report by British neurologist James Purdon Martin and human geneticist Julia Bell on an extended family with 11 mentally retarded males and 2 mildly affected females. None of the affected family members showed any distinguishing physical features in common. Following Penrose's interpretation, however, Martin and Bell regarded this as a rare instance of X-linked intellectual disability, rather than a representative example of a larger trend. The next year, eugenicist-physicians William Allan and C. Nash Herndon of Wake Forest College, along with Allan's secretary, Florence C. Dudley, published another large pedigree suggestive of X-linked mental retardation, in this instance associated with one clinical feature: muscular weakness. Lehrke also pointed to three additional pedigrees published in the early 1960s, one by Monty Losowsky, and two others, already mentioned, by Renpenning and Dunn.[37]

Coupled with the contemporary work of Turner, to whom Lehrke sent a copy of his *American Journal of Mental Deficiency* paper, the number of families in which X-linked mental retardation had been reported was growing quickly in the early 1970s. The increasing acceptance of Lehrke's position on the prevalence of X-linked mental retardation can be inferred from the March of Dimes's interest in publishing an unedited version of his dissertation in 1974, two years after Penrose had died.[38] Indeed, by this time, medical geneticists increasingly understood studies of institutionalized individuals and families with inherited forms of mental retardation as clearly demonstrating that the X chromosome did have a significant role to play in these conditions.

Looking for New Patterns of Malformation

Allan, Herndon, and Dudley, in their 1944 report on inherited mental retardation, divided the genes thought to cause deficiency into three categories: (1) those that caused no other physical signs of malformation, (2) those that did have other discernible effects, and (3) genes causing disorders that sometimes were associated with mental retardation. Among the categories, the researchers noted that the first was likely to have mostly genetic causes, while being the least amenable to further clinical delineation.[39] Three decades later, Gillian Turner sought to accomplish what Allan and colleagues believed they could not, by finding subtle physical differences among individuals with X-linked mental retardation and no major bodily malformations. She had many reasons at this time for believing that further delineation was possible. For one, Herbert Lubs had sent her a copy of his 1969 paper identifying the marker X chromosome, further sparking the idea in her mind that discrete X-linked forms of mental deficiency could be identified. Also, an interesting report of another bodily marker had come from Sao Paulo, Brazil in 1971, where J. A. Escalante, a PhD student, noted that nine males in a family with X-linked mental retardation possessed unusually large testicles (macroorchidism). In addition, the presence of a chromosomal variant similar to the one Lubs had described was also mentioned in Escalante's dissertation, but not in his published report. Turner looked for this X-linked chromosomal marker in the families she was studying but failed to find it. At the same time, she was also on the lookout for instances of macroorchidism among her patients.[40]

The standardized measure of testicle size was first made possible by the development of the orchidometer by Swiss pediatrician Andrea Prader in the mid-1960s. Prader's "orchidometer" comprised multiple egg-shaped pieces of plastic, with specific volumes. A 1974 study of almost 4,000 males established the normal range for testicle growth. In the clinic, testicle size was easily overlooked, especially since abnormal growth often did not occur until adolescence. Soon after Escalante's report, a colleague of Turner's named Chris Eastman, who was studying endocrinology, examined a pair of brothers affected by mental retardation. Using an orchidometer that he had recently acquired, Eastman found the brothers' testicles to be nearly twice the average size, although the brothers had normal bodily levels of testosterone. In 1975 Turner and Eastman reported, in the *Journal of Medical Genetics,* on two pairs of brothers they had identified who were affected by mental retardation, apparently of an X-linked nature, and who seemed to possess only one minor bodily malformation: macroorchidism.[41]

Testicular size thus appeared to be another bodily feature that could be used by medical genetics to further delineate X-linked forms of mental retardation. Other researchers in the United States, Canada, and Mexico offered similar reports of this clinical pattern over the next two years. Pediatrician Rogelio Ruvalcaba and colleagues in Washington state referred to this clinical pattern in a 1977 *Journal of the American Medical Association* article as "X-Linked Mental Deficiency Megalotestes Syndrome," suggesting the existence of a distinct clinical disorder. Similarly, Peter Bowen and his pediatric colleagues in Edmonton distinguished the disorder as "the X-Linked Syndrome of Macroorchidism and Mental Retardation" and alternatively as "the macroorchidism-MR syndrome." These physicians went on to suggest, "The clinical, pathological, and endocrinological findings are sufficiently similar in these and other reported cases to suggest a common pathogenesis and identical cause." Macroorchidism was identified as more than an occasional or variable feature in nonspecific X-linked mental deficiency. By the late 1970s, it was considered a distinguishing bodily marker, representing a distinct clinical pattern.[42]

The link between X-linked mental retardation and unusually large testicles was soon after cataloged in major reference texts. In 1978 McKusick identified "Mental Retardation with Large Testes" as a distinct genetic disorder, giving it a unique entry and number in *MIM*. This pattern of bodily malformation was also included in the third edition of *Recognizable Patterns of Human Malformation* (1982) as "Gillian Turner–Type X-Linked Mental Deficiency Syndrome," a name David Smith coined. This form of X-linked mental retardation had not previously been listed in Smith's text, because it lacked reoccurring bodily malformations. Following the association of mental retardation and macroorchidism, Smith included the disorder among those showing "Early Overgrowth."[43] By the early 1980s, with its entry into these texts, X-linked mental retardation with macroorchidism had become an institutionalized disorder in medical genetics.

Reemergence of a Genetic Marker

Just as accounts of a clinical pattern involving macroorchidism and X-linked mental retardation were published in the mid-1970s, human cytogeneticists began reporting sightings of Herbert Lubs's marker X chromosome, for the first time since 1969. In three papers, published during 1976 and 1977, medical geneticists reported a cytogenetic abnormality similar to the one described by Lubs, in multiple patients with mental retardation. Instead of secondary constrictions, human cytogeneticists in the 1970s referred to these chromosomal findings as "fragile sites." The term had been coined a year after Lubs's 1969 report, and with no reference to his findings, by human cytogeneticists Ellen Magenis and Frederick Hecht at the University of Oregon Medical School. Fragile sites were so named because of the regularity with which chromosomes appeared to break at these locations. Magenis and Hecht described a fragile site on chromosome 16, which did not appear to be associated with any clinical conditions.[44]

During the period between 1969 and 1976, human cytogenetics had transformed from a niche research area to a worldwide practice, thanks in large part to its clinical application in diagnosing Down syndrome and other chromosomal disorders.[45] In the early 1970s, the Children's Hospital in Marseilles, France, in line with many biomedical centers around the world, opened a medical genetics division, under the direction of pediatrician Françis Giraud. Giraud had no genetics training himself, but he hired two clinical geneticists, Ségolène Aymé and Jean-François Mattei, as well as Mattei's wife, Marie-Genevève, a skilled cytogeneticist.[46] In 1976 Giraud's clinical team reported on the identification of a fragile site on the X chromosome in five boys affected by mental retardation, some of whom had unusual facial features. The fragile site was found in the same location on the X chromosome in each case and was similar to the marker X chromosome that Lubs had described seven years earlier. Larger family studies were done on two of these boys, revealing the fragile X site in their intellectually normal mothers as well. The next year, Jill Harvey, a human cytogeneticist in Australia, reported on four additional families with 20 mentally retarded males possessing the fragile X site. All of the males that Harvey and colleagues examined were normal in appearance.[47]

Why had it taken almost a decade for Lubs's 1969 finding to be independently verified? Medical geneticists around the world had been aware of Lubs's report and were actively looking for similar marker X chromosomes.[48] Reflecting on the years that had gone by without another marker X being reported, Harvey and colleagues pointed, as Lubs had in his own report, to the variable expression of the fragile X site. In most individuals the marker X chromosome was seen in fewer than 30% of examined cells. This factor, coupled with the subtlety of the marker, made it easy to miss. Harvey emphasized that a cytogeneticist looking for the fragile X site should analyze 100 cells, making note of any minor chromosomal abnormalities.[49]

In 1977 another Australian cytogeneticist, Grant Sutherland, offered additional insight on the question of how the fragile X site had gone unseen for almost a decade. Sutherland had been among those looking for new cases of mental retardation showing the fragile X site. Working in Adelaide, Australia, during the mid-1970s, Sutherland too had had no success identifying the fragile site in affected families.[50] Frustrated, he visited Jill Harvey in Melbourne and found that her laboratory, where the fragile X site had been identified in multiple families, still used culture media 199 for growing cells. Sutherland later referred to 199 as "the Model T Ford of tissue culture media." Most laboratories, including his in Adelaide, had switched to a newer media, like Ham's F10, years earlier. Returning to Adelaide, Sutherland found success making the fragile X site visible in affected families when he used the older cell culture media. From here, Sutherland sought to figure out what made media 199 different. Ultimately, he determined that it was not what the media had in it that made it useful for fragile site detection, but rather what it lacked: specifically, folic acid. A higher concentration of folic acid was added to newer media, which most laboratories had adopted around 1970, in order to enhance cell growth. This change seemed to explain the inability of laboratories to make fragile sites visible around this time.[51]

Human cytogeneticists across the world who were interested in identifying fragile sites quickly adopted Sutherland's protocol. Art Daniel, a cytogeneticist working with Gillian Turner in Sydney, continued to experience difficulty in eliciting fragile site expression. However, after consulting with Sutherland and adjusting the pH of his culture medium, Daniel was successful in making the fragile X site visible in some of Turner's patients. Among 18 families with X-linked mental retardation and no major malformations that were reexamined, 8 showed the fragile X site. Even more remarkably, these 8 families were the only ones in which affected males also had usually large testicles. Thus, the pattern of malformation involving X-linked mental retardation and macroorchidism appeared to also include Lubs's fragile X site.[52]

Earlier in the 1970s, dysmorphologists had interpreted the association between X-linked mental retardation and macroorchidism as suggesting a clinically distinct disorder, which likely had one genetic cause. Using the fragile X site as a distinguishing marker among families with X-linked mental retardation and no major malformations, medical geneticists also began to identify other features common to this population, including large ears, a long face, and a prominent forehead.[53] Indeed, as McKusick suggested in his introductory section of *MIM*, the identification of genetic markers, such as the fragile X site, allowed researchers to examine particular disorders "in pure culture."[54] This made medical geneticists able to distinguish clinical features common to patients with a particular genetic disorder from normal bodily variability among individuals.

As sociologist Adam Hedgecoe has described, the identification of a genetic marker for a disorder often brought about the recognition of new bodily characteristics and facilitated the diagnosis of patients with more variable clinical expressions, leading to phenotypic "expansion and uncertainty."[55] During the 1980s, clinicians developed a more nuanced understanding of the effects that possessing the fragile X site had on females as well, some of whom showed learning disabilities. The introduction of a genetic marker, here as in other cases, reshaped expectations about the disorder's life history and patterns of malformation.[56]

The fragile X site also shaped assumptions in medical genetics about the nature of visible chromosomal abnormalities. Disorders inherited in a Mendelian X-linked recessive pattern were not expected to have a visible genetic marker, because medical geneticists assumed that they were caused by a DNA-level gene mutation. During the early 1970s, human cytogeneticists were often frustrated when physicians sent them samples for chromosomal analysis from patients who clearly had Mendelian disorders. McGill University cytogeneticist F. Clarke Fraser recalled that "there was something of an overreaction; genetics was regarded by some as synonymous with cytogenetics, genetic diseases with chromosomal diseases." As historian Ilana Löwy has insightfully argued, however, many medical geneticists were happy to see their jurisdiction of genetic diseases being expanded to include chromosomal disorders at this time. Indeed, the field benefited significantly from the growing uptake of prenatal diagnosis of Down syndrome.[57]

While definitions were evolving during the 1970s, the fragile X site did blur the lines that traditionally distinguished genetic (hereditary) from chromosomal disorders. In 1978 McKusick noted in his *MIM* entry for

"Mental Deficiency (Martin-Bell or Renpenning Type)" that the inheritance pattern of the fragile X site suggested that chromosomal analysis could in fact aid in the diagnosis of other single-gene Mendelian disorders.[58] As mentioned previously, the fragile X site was a point of overlap between two traditionally distinct categories of genetic disease: Mendelian disorders, caused by Penrose's "mistakes of an imaginary printer," which were cataloged in *MIM*, and the chromosomal syndromes featured in *Recognizable Patterns of Human Malformation*, which resulted from the "mistakes of a binder."[59] The fragile X site thus offered an important example, in the late 1970s, of the potential for the genetic causes of clinical disorders to be identified and assessed, based on visual chromosomal analysis.

Preventing X-Linked Mental Retardation

Beginning with Herbert Lubs's report in 1969, and in many reviews of medical genetics published in the decades thereafter, the fragile X site was pointed to as an example of how genetic disorders could be delineated based on visible chromosomal abnormalities. Researchers characterized the fragile X site, and other genetic markers like it, as representative of the potential for targeted prevention.[60] In a 1978 *New England Journal of Medicine* report, Gillian Turner and colleagues described how physicians might soon make use of the fragile X site. When a male with X-linked mental retardation and macroorchidism was identified, his chromosomes would be examined to detect the fragile site. If it was found, other family members could be examined and female carriers identified if the fragile X site was made visible in their cells. Based on the expected inheritance pattern for Mendelian recessive X-linked disorders, medical geneticists believed that if a woman had an affected brother, she had a 50% chance of being a carrier. If she was a carrier, this meant a 50% chance of having an affected son. Based on carrier status, prenatal diagnosis using amniocentesis could be performed; but as with Down syndrome, no treatment existed beyond targeted prevention through abortion.[61]

In addition to identifying female carriers in fragile X families, cytogenetic screening created opportunities to identify new families with the fragile X site, in which the disorder had not yet been clinically expressed or diagnosed. Based on the finding that carrier females often showed minor learning disabilities, medical geneticists began to talk about the disorder as being X-linked dominant, rather than recessive. Turner and colleagues surveyed girls throughout New South Wales, Australia, who received special education because of learning disabilities, for the fragile X site. The study identified 5 girls, out of 72 tested, who showed this genetic marker. Based on these findings, families could be surveyed for additional female carriers, who would be counseled that they had a 50% chance of passing on the trait, likely resulting in either a mentally deficient son or a more mildly affected daughter, who would also be a carrier.[62]

As Ruth Schwartz Cowan has elegantly described, during the 1970s and 1980s, a technological system for prenatal diagnosis and prevention was slowly assembled in medical genetics.[63] The system primarily targeted Down syndrome but was also applied to other genetic disorders, including X-linked mental retardation and macroorchidism. In the late 1970s, it was unclear to medical geneticists whether the fragile X site would appear in all carrier females, or whether it could be made visible when tested for prenatally. Establishing the fragile X site as a reliable indicator of risk for mentally retarded offspring required additional cytogenetic observation and experience. Reports from Australia and the United States, published during 1979 and 1980, verified the presence of the fragile X site in a subset of mentally retarded males. The visibility of this cytogenetic abnormality in their mothers, however, proved to be much less reliable. Sutherland reported identifying the fragile X site in only 5 of 13 women who had mentally retarded sons with the marker. Cytogeneticist Patricia Jacobs, in Hawaii, was able to identify the marker in only 6 of 19 mothers. Similarly, Patricia Howard-Peebles reported seeing the fragile X site in just one of six mothers of boys who had the disorder.[64] Based on Mendelian inheritance patterns, medical geneticists assumed that the fragile X trait was present in these women, but they were not always able to make it visible.

Offering prenatal diagnosis to at-risk women using amniocentesis also required that the fragile X site could be made visible among the fetal skin cells present in amniotic fluid. In 1979 Sutherland reported that he had not been able to identify the fragile X marker in the skin cells of affected individuals. The next year, medical geneticists Peter Jacky and Frederick Dill, in Vancouver, demonstrated that small variations in how chromosomes were handled brought about the expression of fragile sites in skin cells. Not long after, Edmund Jenkins, a cytogeneticist in New York, reported that chemical manipulation with fluorodeoxyuridine (FUdR), a common chemotherapy agent, made the fragile X site visible in amniotic cells. Jenkins and his colleagues had conducted a retrospective study on prenatal samples from individuals clinically diagnosed with the fragile X form of mental retardation. After using various different chemicals, the researchers eventually hit on FUdR as a reliable means of inducing fragile X expression. With these prenatal cytogenetic techniques in place, Jenkins and others began to study the feasibility of offering prenatal diagnosis and prevention to fragile X families.[65] Many cytogeneticists had already noted, however, that the fragile X site often did not behave as expected. This posed significant challenges for medical geneticists and delayed the uptake of prenatal prevention.

As efforts to develop prenatal diagnostic testing moved forward in the early 1980s, some medical geneticists sought to expand clinical awareness of the disorder. At this time, human cytogeneticist Loris McGavran and pediatrician Randi Hagerman at the University of Colorado began looking for the fragile X site in their own mentally retarded patients. Hagerman had become interested in the disorder after becoming aware of a local patient with mental retardation and large testicles. She recruited McGavran, who succeeded in making visible the fragile X site in the boy's cells. Over 18 months, the two collaborated to identify 25 more fragile X patients.[66] Hagerman was disappointed by the lack of publications on the disorder, and so in 1983 she and her colleagues decided to put together a book, which they titled *The Fragile X Syndrome: Diagnosis, Biochemistry, and Intervention*. At the time, many different names were in use. In the 1983 edition of *MIM*, McKusick referenced several of them: "X-Linked Mental Retardation and Macroorchidism; Marker X syndrome; Fragile X syndrome; Martin-Bell Syndrome."[67] With the publication of their book, Hagerman and colleagues sought to establish a single name for the disorder, based on the presence of the fragile X site, which they found to be its most striking feature.[68] While other names continued to be used throughout the 1980s, the designation *fragile X syndrome* came to define this form of mental retardation, as it increasingly became a target for genetic diagnosis and prevention using amniocentesis and abortion.

Unruly Behavior

Even though the fragile X site offered the potential opportunity to diagnose and prevent one form of inherited mental retardation, medical geneticists continually found that this chromosomal marker did not

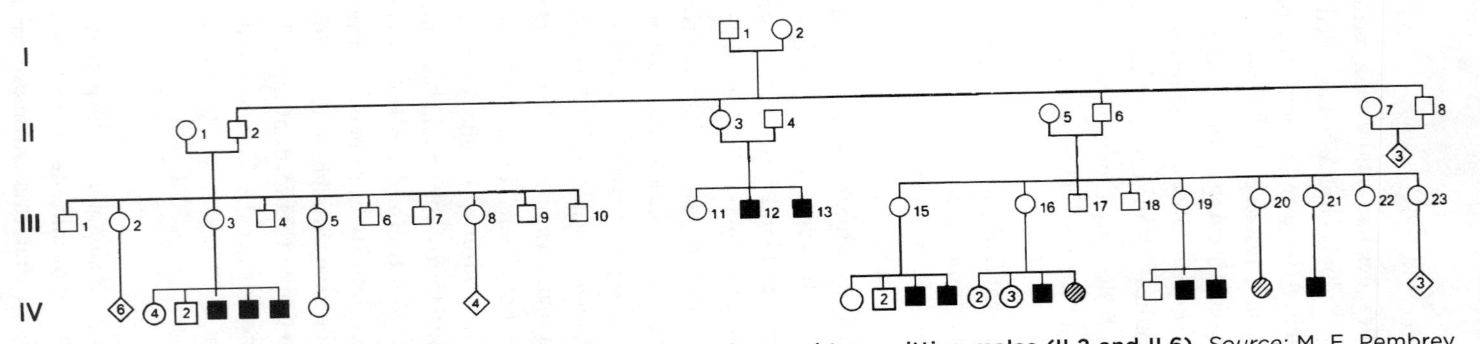

Pedigree from a fragile X family demonstrating the existence of normal transmitting males (II.2 and II.6). *Source:* M. E. Pembrey et al., "A Premutation That Generates a Defect at Crossing Over Explains the Inheritance of Fragile X Mental Retardation," *American Journal of Medical Genetics* 21, no. 4 (1985): 711. Reprinted with permission from John Wiley and Sons. Copyright 1985 Alan R. Liss, Inc.

behave like a normal X-linked Mendelian trait. In addition to not being visible in all, or even most, bodily cells of affected individuals, the variant also often could not be made visible in female carriers. As more affected families were identified by way of cytogenetic examination in the early 1980s, pedigree studies were conducted. Based on the results, medical geneticists were surprised and unnerved to find that in many instances mental retardation had suddenly appeared in multiple branches of these families without any warning.[69]

Normally, when multiple cousins across an extended family were found to have the same rare genetic disorder, medical geneticists assumed that a grandparent or great-grandparent must have also possessed the trait and passed it down to them. In the case of X-linked traits, the disorder might skip a generation or two when it was passed down exclusively through females. In fragile X syndrome family pedigrees, however, researchers regularly came across unexpected instances in which multiple brothers each had affected grandchildren. Based on their assumption that this X-linked trait would affect all males who possessed it, these brothers should have been affected by fragile X syndrome themselves. In many families, however, neither bodily nor chromosomal features of the disorder were detected in males who had transmitted the trait.[70]

With an X-linked model for the inheritance pattern of fragile X syndrome in mind, medical geneticists initially discounted some of the experiences that affected families were reporting to them. Since researchers believed that all males who passed on the X-linked trait must also be affected by it, they ignored the paternal pedigrees of female carriers when their father was unaffected. Biostatistician Stephanie Sherman later recalled that women with affected children would often talk about mental retardation on their father's side of the family, only to be told that this was irrelevant to fragile X syndrome unless a woman's father was affected. As Sherman later acknowledged, "we didn't listen to the families in the beginning." Eventually, however, the combination of many pedigrees in which unaffected males seemed to be passing down the disorder and family accounts of mental retardation on a father's side persuaded medical geneticists to reconsider their assumptions. As historian Susan Lindee has lucidly demonstrated, on-the-ground family experiences are often central to the production of new genetic knowledge.[71]

As part of her research on the disorder, Sherman analyzed a large number of pedigrees from fragile X families and found that only about 40% of the sons of female carriers, instead of the expected 50%, showed clinical signs of the disorder. The fragile X trait was frequently being passed down through males whom she called "normal transmitting males," who theoretically should have been affected by the disorder because they had passed it on to their mentally retarded grandchildren. Instead, in the pedigrees that medical geneticists collected from these families, the incidence of fragile X syndrome appeared to increase over the generations, suggesting an instance of the controversial pattern of inheritance called "anticipation."[72]

Like X-linked mental retardation, anticipation was not widely accepted in the genetics community because of the opposition of Lionel Penrose. Penrose believed, and in the minds of most geneticists had convincingly demonstrated, that anticipation was an artifact of genetic data bias rather than a biological phenomenon. Because medical geneticists were much more likely to see affected individuals in a family than healthy ones, there was likely to be an "ascertainment bias" in their findings. This, Penrose argued, was what made researchers erroneously identify "anticipation," which was an apparent increase in the severity of a disorder over the generations. Because of Penrose's opposition, medical geneticists who were examining fragile X pedigrees were hesitant to suggest anticipation as an explanation for the normal transmitting males they observed.[73]

John Opitz, an influential physician and longtime editor of the *American Journal of Medical Genetics,* was particularly struck by Sherman's findings and began referring to her identification of apparent anticipation in fragile X families as "the Sherman Paradox." Sherman, a young researcher at the time, was somewhat uncomfortable about being associated with these findings. She later recounted, "I was very concerned that what I was reporting was going to lead me to a non-scientific career. . . . In the back of my mind I kept thinking this [Sherman's Paradox] is going to be Sherman's folly." To argue against Penrose (who had died more than a decade earlier) and the established genetic dogma that he represented was not a safe career move.[74]

From a basic research perspective, the unusual inheritance pattern of the fragile X site offered a unique and exciting (if somewhat risky) opportunity for medical genetics theorizing. Opitz reflected on the unusual inheritance pattern in his preface to the 1986 report of the Second International Workshop on the Fragile X and X-linked Mental Retardation. Did a single inherited gene cause this disorder, or was the fragile X site itself mechanistically to blame? Opitz leaned toward the belief that an "unconventional" genetic mechanism causing fragile X syndrome would be found, "one that is perhaps prototypic of a whole new class of disorders intermediate between Mendelian and aneuploidy [chromosomal] syndromes."[75] For Opitz, fragile X syndrome seemed to bridge the gap between two distinct types of disease etiology in medical genetics.

Other participants in the Second International Workshop offered many hypotheses concerning the genetic mechanism behind fragile X syndrome. Some suggested that a transposable piece of genetic material was passed down through unaffected generations and eventually inserted into the X chromosome, bringing about the fragile site and the disorder.[76] Grant Sutherland offered an amplification model, in which particular strands of DNA repeats were made longer through errors in replication and over time formed a causative lesion.[77] Along these lines, molecular geneticist Robert Nussbaum suggested a mechanism by which recombination increased the length of DNA repeats, interfering with a gene and thus causing fragile X syndrome.[78] Marcus Pembrey and Kay Davies drew on an earlier model suggested by Opitz (for the inheritance of achondroplasia), of an "unstable premutation" in forming their hypothesis. This premutation itself was thought to have no clinical effects but was predicted to later transform into a causative mutation, when passed from a female carrier to her son.[79]

Some of these models were actively being tested in the laboratory during the mid-1980s, as cytogenetic analysis of the fragile X site began to incorporate molecular techniques, along with much of medical genetics at the time.[80] To eventually prove a particular explanation, researchers believed that the genomic region around the fragile X site would have to be molecularly located, isolated, and sequenced at the DNA level. This was representative of a larger expectation of late-twentieth-century medical genetics, that molecular analysis offered the opportunity to achieve a more objective view of genetic mutations and how they cause disease. For medical geneticists, their experience with the fragile X site further demonstrated that cytogenetics was messy, subjective work. Chromosomes often did not behave as anticipated, leading to results that conflicted with Mendelian understandings of disease inheritance. This undercut medical confidence in the reliability of one mutation–one disorder explanations, upon which prenatal diagnosis and targeted prevention depended.

A Molecular Model for Fragile X Syndrome

Throughout the late 1980s and early 1990s, researchers attempted to locate a molecular-level gene, mutation, or abnormality that could account for the chromosomal presence of the fragile X site, as well as the clinical expression of fragile X syndrome.[81] In 1991 the location of the causative genetic trait for fragile X syndrome was identified and found to be in close proximity to the fragile site. Representative of research on X-linked mental retardation throughout the previous two decades, this was an international accomplishment, involving contributions from teams in Australia, the Netherlands, France, and the United States.[82]

Anne Vincent, Jean-Louis Mandel, and colleagues, in Strasbourg, France, offered evidence that the fragile X region was abnormally "methylated" in affected individuals (more on this below), while an Australian team molecularly demonstrated the instability of the fragile X site, showing that it sometimes grew significantly in size when passed down from mother to child. Annemieke Verkeck and Ben Oostra in the Netherlands, working with American colleagues including Stephen Warren, were able to sequence the DNA in this region, finding that it contained a gene, which the researchers named FMR-1 (fragile X mental retardation–1), accompanied by a long string of CGG trinucleotide repeats. These researchers also found that the number of consecutive CGG repeats in the FMR-1 area was directly related to clinical outcome in fragile X families. Normal individuals appeared to have fewer than 52 CGG repeats in the fragile X region, while those with a "premutation" had between 52 and 200. Premutations were at risk for expanding further when passed from mother to child, with fragile X–affected children having 200 or more CGG repeats—a "full mutation."[83] Functionally, full mutations were characterized by their direct affect on FMR-1 gene activity. When 200 or more CGG repeats were present, this region often became "methylated." Methylation involved the addition of small molecules onto certain DNA nucleotides in the region, which in effect turned "off" the FMR-1 gene. The lack of FMR-1 gene product seemed to cause fragile X syndrome.[84]

By the mid-1990s, once new molecular techniques for identifying premutations and full mutations were clinically proved, the troubled era of

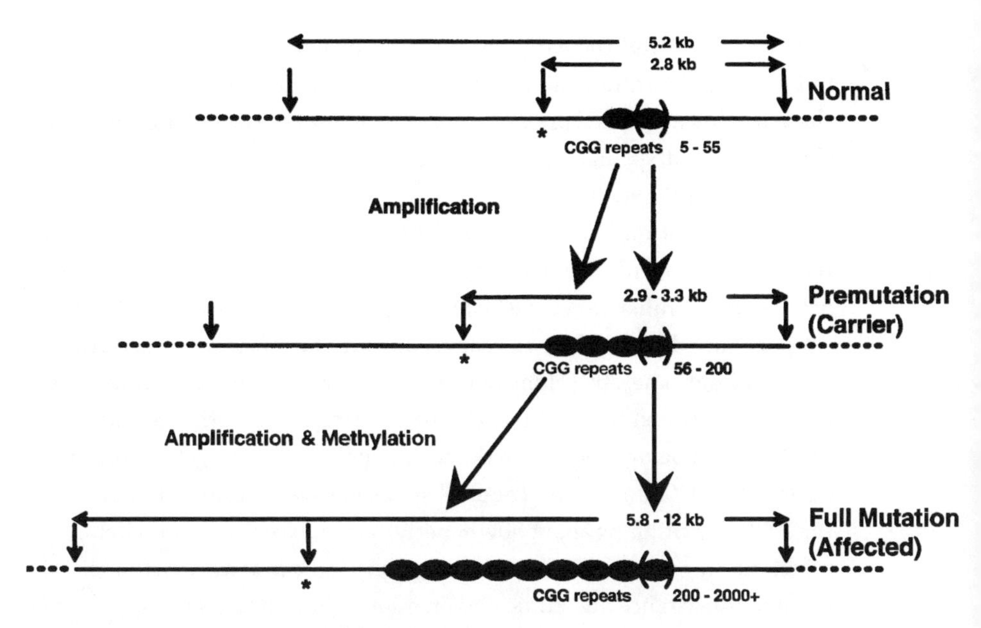

Molecular-level depiction of a fragile X syndrome premutation and full mutation, showing the different number of CGG repeats seen in each. *Source:* Randi Jenssen Hagerman and Amy Cronister, eds. *Fragile X Syndrome: Diagnosis, Treatment, and Research,* 2nd ed. (Baltimore: Johns Hopkins University Press), 95, fig. 2.2. © 1996 The Johns Hopkins University Press. Reprinted with permission of Johns Hopkins University Press.

often misleading cytogenetic diagnosis for fragile X syndrome came to an end.[85] Molecular analysis improved the identification of carriers of the fragile X trait and offered greater certainty for the purposes of prenatal diagnosis and targeted prevention. Uncertainty about the inheritance pattern of fragile X syndrome was explained and resolved through the corollary of an unusual mechanism, but one that was in keeping with Mendelian understandings of genetic disease. The one (full) mutation–one disorder ideal was maintained in the context of fragile X syndrome, meaning that reliable prenatal diagnosis and targeted prevention through abortion could be offered to affected families.

Notably, medical geneticists soon recognized, once again with the assistance of experiences from affected families, that fragile X premutation carriers were themselves often affected by certain clinical disorders. Females with a fragile X premutation were likely to experience premature ovarian failure and the early onset of menopause. Premutation males often developed Parkinsonian tremors and ataxia late in life. As a result of these findings, fragile X premutation status, beyond the risk for intellectually disabled offspring, also became an important indicator of other significant clinical outcomes.[86]

The association of fragile X syndrome with an expanded trinucleotide repeat also led medical geneticists to hypothesize that other clinical disorders that had inheritance patterns suggesting anticipation might have a similar cause. During the two years following the molecular description of fragile X syndrome, both Huntington's disease and myotonic dystrophy were also demonstrated to involve the expansion of trinucleotide repeats.[87] Indeed, beyond being an early model for the promises of targeted prevention during the 1970s and 1980s, fragile X syndrome also became an exemplar in the 1990s for thinking about how genetic mutations with complex behaviors might play a role in other disorders that showed similarly unorthodox patterns of inheritance and clinical expression.[88] Findings from fragile X syndrome led research geneticists to think about genomic function in entirely new ways, an outcome that is representative of the regular and important contributions of clinical practice to basic biology, and of the constant circulation of genetic markers and knowledge between laboratory and clinic.[89]

Conclusion

The life history of fragile X syndrome presented here offers a representative model of how medical geneticists integrated different streams of clinical and cytogenetic knowledge and experience during the late twentieth century in order to delineate discrete genetic disorders. Central to this history was the development of new ways of seeing and interpreting patterns of bodily malformation and chromosomal abnormality. In the 1970s the subspecialties of dysmorphology and human cytogenetics worked synergistically to enhance the diagnosis of genetic disease through the establishment of one mutation–one disorder associations. Fragile X syndrome served as a compelling exemplar of this model and its promises. In spite of decades of uncertainty, medical geneticists demonstrated that the combined efforts of cytogenetic and clinical analysis could facilitate the prevention of mental deficiency through prenatal diagnosis and abortion.

Reflecting this enthusiasm, in a 1977 National Institutes of Health report, physicians Herbert Lubs and Felix de la Cruz argued that medical genetics had entered the "age of prevention," noting that because curing genetic disease was much more difficult than preventing it, treatment was coming to be understood as a secondary aim.[90] Of course, the goal of preventing disease was nothing new for medical geneticists. Indeed, the late-1970s "age of prevention" represented another iteration of what Nathaniel Comfort termed the eugenic impulse.[91] But this version of the impulse was different from past iterations. During the 1970s medical geneticists felt much more confident in their ability to target particular disorders for prevention than they had three decades earlier. While J. B. S. Haldane lamented the lack of tools for prevention in 1949, midcentury medical geneticists were much closer to this aim than da Vinci had ever been to flight. Throughout this period, and despite various complications, fragile X syndrome was used as a model for how genetic testing could prevent mental retardation.[92]

But what about a cure? Victor McKusick, the most prominent leader of late-twentieth-century medical genetics, smartly realized that public funding was likely to be limited if researchers could offer nothing more than prenatal diagnosis and prevention through abortion. In his writing, McKusick drew upon the distant dream of gene therapy, finding and replacing mutations with normal genes, to promote sustained private and government investment in identifying the genetic basis of disease. Looking forward, McKusick confidently stated that locating the causes of more genetic disorders among the human chromosomes would certainly enhance the future scope of potential genetic treatments. While gene therapy remained on the horizon, McKusick and other medical geneticists continued to promote genetic analysis as a means for expanding the reach of targeted diagnosis and prevention. In the decades to come, human gene and disease mapping produced many new targets for prevention, but still few treatments.[93] The next chapter explores how advances in human cytogenetics during the 1970s and 1980s contributed to gene and disease mapping and the construction of a chromosomal infrastructure in medical genetics for diagnosis and prevention.

Chapter 2

Chromosomal Cartography

In September 1969 Victor McKusick addressed the Third International Conference on Congenital Malformations, held in the Netherlands. McKusick's talk, given just two months after the successful American moon landing, envisioned another major government investment in the exploration of a largely unknown territory, the human chromosomes: "An exhilarated moon-landing state of mind leads us, perhaps properly, to consider what areas of human biology might be concentrated on in an all-out, maximally supported attack. . . . The chromosomes of man are still largely terra incognita. The developments in human cytogenetics in the last 10–15 years have shown us the gross outlines of the continents. . . . In a pitifully small number of instances we know pairs of neighbors [human genes] residing on one of the continents." His argument drew on space-age excitement among the public, as well as government interest in funding exploration projects that demonstrated great technological capabilities and promised the discovery of new and useful knowledge. McKusick's presentation, entitled "Birth Defects—Prospects for Progress," put forth mapping the human chromosomes as such an undertaking and as an important investment in long-term efforts to reduce the incidence of inborn disorders. While he readily admitted that the likely contributions of human gene mapping to prevention were poorly defined, McKusick expressed confidence that all birth defects had some genetic basis.[1]

Through his advocacy, McKusick was positioning medical genetics to benefit from an era of unprecedented public expenditures supporting science and medicine. Following World War II, and in the midst of Cold War competition with the Soviet Union, the US government invested heavily in basic research, including various "big science" projects with lofty goals such as putting a man on the moon.[2] McKusick pitched mapping the human chromosomes as another massive and worthy government investment, one that was sure to offer a significant return in the decades ahead. In 1971 McKusick predicted that mapping all of the human genes was a project that might take more than a century to complete. Later in the decade, however, he began pointing to the year 2000 as a target date, a conjecture that proved accurate.[3]

McKusick gave his "Birth Defects—Prospects for Progress" speech in the same year in which Herbert Lubs published his report of the marker X chromosome and its association with inherited mental deficiency. Lubs's paper, described in chapter 1, reflected the many promises and challenges that human cytogeneticists faced in the late 1960s. While they were able to make abnormalities associated with disease visible, it remained difficult to determine on which chromosomal "continent" these mutations were located. This situation was demonstrative of the reality that in 1969 human cytogeneticists had no infrastructure upon which to map genes and mutations: they lacked a set of standardized landmarks along the chromosomes to use in describing where abnormalities had been identified. As a result, it was nearly impossible for cytogeneticists to determine whether the fragile site they detected in one patient was the same site that had been previously reported in another. Given this situation, the goal of mapping all human genes and disorders along the chromosomes, and thus facilitating the prevention of birth defects, appeared to be as distant as the moon landing had a decade before it occurred.

Central to the eventual successes of human gene mapping were collaborative efforts to enhance and standardize the look and geography of human chromosomes. During the 1970s human cytogeneticists developed the basic chromosomal infrastructure upon which human genes continued to be mapped four decades later. Though human gene mapping eventually became a molecular and highly computerized process, in the 1970s it involved a slow and haphazard process of isolating chromosomal fragments within living cells to determine whether or not certain genes were situated at specific chromosomal locations. The establishment of a repository for useful genetic variants, international research workshops that were occurring, and the continuous bidirectional exchange of knowledge and resources between laboratory and clinical spaces contributed significantly to quickening the pace of early human gene mapping projects.

This chapter describes the genetic and clinical techniques, funding sources, and international collaborations during the 1970s that made the completion of a human gene map seem feasible before the close of the twentieth century. McKusick and other influential medical geneticists had a significant role in making the case for human gene mapping among their fellow scientists and physicians, as well as potential funders and the broader public. These individuals associated gene mapping with the exploration of distant lands, great intellectual achievements, improved understandings of human evolution, efforts to reduce the incidence of disease, and advances in treating disorders. The human chromosomes were presented as a new frontier for scientific and medical research that was certain to be a gold mine for better understanding and improving the human condition.

By the late 1960s, the "Golden Age" of generous US government funding for collecting scientific knowledge of undetermined future value was nearing its end.[4] Nonetheless, to buoy their undertaking, human gene mappers continued to rely heavily on the persuasive power of the already fading space-age technological optimism. Indeed, with little in the way of immediate practical applications for gene mapping to tout, McKusick and his colleagues had few options other than to depend on the willingness of their funders, public and private, to believe that investments in human gene mapping would undoubtedly bear fruit in the decades to come. They relied on visual depictions of human gene mapping in progress, and drew on cartographic and anatomical references, to convey the success, future promise, and widespread scientific and medical relevance of human gene mapping. Another key aspect of the long-term success of human gene mapping was to maintain and increase momentum. By constantly highlighting what had already been accomplished and what was still left to achieve, advocates of gene mapping were able to excite other researchers and potential funders about what a completed map would mean for the prevention of disease and improvement of public health.

The expansion of human gene mapping during the 1970s and 1980s added a new dimension to the life history of many disorders. Closely related to the identification of human genes was the localization of clinical disorders and their genetic causes along the chromosomes. As medical geneticist Reed Pyeritz, a student of McKusick's during this time, later recounted, "When people started laying out the 23 sets of chromosomes, the ideograms, and next to [them] where a gene had been identified, [McKusick] said, 'that's all well and good, but you can often map a [disease] phenotype to a specific site on a chromosome before you know what the cause is.'"[5] Mapping genes and disease phenotypes represented two sides of the same coin. One could map a normally functioning gene and ponder the potential effects of mutation, or map a pathological phenotype in the hope of locating the gene with which it was associated. In the late twentieth century, medical geneticists worked in both directions to expand their genetic knowledge. These ongoing efforts to map disorders along the human chromosomes reflected and reinforced the one mutation–one disorder ideal of postwar medical genetics.

Building an Infrastructure

The late 1950s to the early 1960s was a time of significant productivity for human cytogenetics. The establishment of the human chromosome number in 1956 led in the ensuing years to the identification of various abnormal chromosomal counts that were associated with specific clinical disorders, including Down, Turner, and Kleinfelter syndromes. As researchers began to associate clinical disorders with specific chromosomes, they agreed upon the need for a standardized naming system. In 1960 cytogeneticist Theodore Puck organized an international meeting in Denver to standardize chromosomal nomenclature. A small group of cytogeneticists who had already published research on human chromosomes were invited to the conference.[6] Like many meetings on nomenclature, the discussions were at times contentious. Ultimately, however, the group decided to name the human sex chromosomes X and Y and to number the 22 others (autosomes) by their relative size, with chromosome 1 being the largest.

The nomenclature that the meeting participants developed in 1960 continued to be used more than 50 years later in the construction of karyotypes. Human cytogeneticists began karyotype construction by photographing all the chromosomes in a human cell. Each chromosome was then cut out of this image and arranged by size, often into four rows. Although each chromosome was given a number in 1960, human cytogeneticists lacked sufficient visual markers to reliably distinguish many individual chromosomes. Because of this, cytogeneticists also divided the human chromosomes into seven groups (A–G) that could be visibly differentiated from one another. The largest, the C group, contained eight chromosomes (6–12 and X), which were too similar in size and shape to be distinguished. Human cytogeneticists depicted the defining characteristics of these groups with a set of ideograms, idealized representations of each chromosome's structure.[7]

After the Denver meeting, a core contingent of human cytogeneticists scheduled regular follow-up nomenclature meetings that coincided with

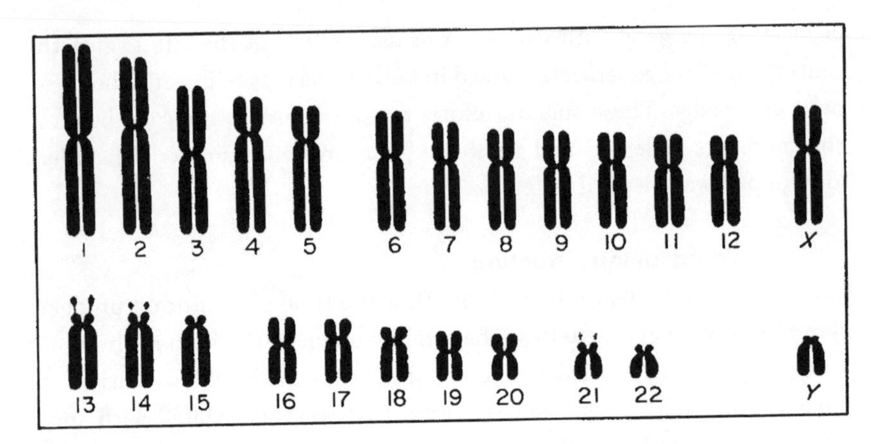

Set of chromosomal ideograms from the 1960 Denver Conference on Human Cytogenetic Standardization. *Source:* Editorial comment, "A Proposed Standard System of Nomenclature of Human Mitotic Chromosomes (Denver, Colorado)," *Annals of Human Genetics* 24 (1960): 319. Reprinted with permission from John Wiley and Sons. Copyright 1960 John Wiley and Sons Ltd./University College London.

the International Congress of Human Genetics, which was held every five years. In 1966, during the congress's meeting in Chicago, a conference was convened to update cytogenetic nomenclature.[8] One of the most significant outcomes of this gathering was the designation of abbreviations for distinct chromosome parts. Each human chromosome had previously been divided at its centromere into a long arm and a short arm. At the nomenclature meeting, it was agreed that the short arm of each chromosome should be abbreviated *p* for *petit*. As American cytogeneticist Kurt Hirschhorn recalls it, the debate over how to designate the long arm was extended and contentious. Participants at this and other nomenclature meetings later suggested that many of the disagreements took place along national lines, particularly between French and German members.[9]

The long-arm naming debate lasted late into the night, in large part because a francophone designation had already been given to the short arms. As Hirschhorn tells the story, sometime after midnight, Lionel Penrose, that year's congress president, walked into the room where the meeting was being held and was surprised to find that the discussion was still going on. Informed that the designation *p* had been decided upon for the short arm of chromosomes, Penrose immediately suggested that the long arm should be called *q*. This was not because *q* had any sort of linguistic significance, but instead because p+q=1 was a well known equation (named after Hardy and Weinberg) in population genetics. Hirschhorn paraphrased Penrose's successful argument in this way, "If you have 'p' for the short arm, use 'q' for the long arm: 'p+q=1,' you got the whole chromosome."[10] Apparently Penrose's suggestion settled the debate. This 1966 decision was just one of the many hard-fought compromises that occurred in ongoing international discussions about the geography of human chromosomes. As in any other mapping project, political and social alliances played a significant role in debates over the identification, naming, and characterization of important landmarks.

In addition to the challenge of agreeing on a common nomenclature, the field was also coming up against the visual limitations of its chromosomal techniques in the 1960s. As we saw in chapter 1, delineation was an issue for Herbert Lubs in his studies of the marker X chromosome. He inferred that the abnormality he saw was on the X chromosome, based on the inheritance pattern of mental retardation in the family he studied.[11] The lack of standardized features for visually distinguishing chromosomes was a major problem for human cytogeneticists as they sought to identify and characterize this and other reoccurring abnormalities. Two researchers in different locations might believe that they were seeing the same mutation under the microscope, but they had no reliable means of knowing that it was on the same chromosome. Reflecting on these issues, cytogeneticist Margery Shaw referred to the late 1960s as an era when human cytogenetics was "in the doldrums."[12]

For any sort of gene mapping to proceed, human cytogeneticists needed to better define the chromosomal landscape. Geneticists had long believed that certain genes were regularly inherited together because they were physically linked on the same chromosome, but this did not aid in mapping genes to specific locations if the chromosome on which they were colocated could not be determined. To use a cartographic metaphor, assigning Mt. Kilimanjaro and Lake Victoria to the same continent would be of limited value to mapmakers if they could not distinguish Africa, South America, and Australia. Similarly, with no visual means to differentiate many of the human chromosomes, geneticists could identify genes that were physically inherited together but could not associate these genes with specific chromosomal locations.

A major breakthrough came with the adoption of new staining techniques in human cytogenetics, which made reproducible banding patterns visible along each chromosome. Staining chromosomes had always been necessary to identify them under the microscope, but the stains formerly used by human cytogeneticists made all of the chromosomes appear in the same solid color. Quinicrine (Q) banding, developed by Swedish researchers in the late 1960s, relied on fluorescent mustard to reveal distinguishing bands along each chromosome. Giemsa (G) and reverse (R) banding, introduced soon after, also produced patterns along the chromosomes, which were easier to visualize under the microscope. These stains provided cytogeneticists with visible markers that could be used to reliably distinguish each human chromosome.[13] Notably, the bands were also adopted as valuable landmarks for identifying discrete locations along each chromosome, and these were used to physically map human genes and genetic disorders.

Following the 1971 meeting of the International Congress of Human Genetics in Paris, a group of human cytogeneticists gathered in a hotel near the Orly Airport to design a new set of standardized banded ideograms of each human chromosome. The meeting was beset once again by international tensions, with French participants preferring R banding instead of the more widely used G banding. G and R banding produced the same pattern, but in reverse: G bands were light where R bands were dark. The point of disagreement was over which bands should count. Those who did R-banding wanted to number the visible bands down each chromosome as 1, 2, 3. G-banding proponents wanted the same system, but with the dark bands that they saw. A breakthrough came, according to participant Dorothy Warburton, when committee chairperson John Evans suggested a compromise: chromosomes should not be thought of as having regions that were either banded or not. Instead, the bands on each should be understood as continuous: light, then dark, then light again. This way, a standard numbering system could be adopted, with the darkness or lightness of a specific band depending on which staining method was used.[14]

The 1971 committee members decided that the long and short arms on each chromosome should be divided into between one and four regions, which were then further separated into individual bands. Each band was identified by the chromosome (1–22, X, Y) and the arm (*p* or *q*) on which it was located, followed by a number identifying the region (beginning with

1 at the centromere), and finally with a second number to designate the specific band. For example, the band 15q12 (pronounced 15, *q*, 1, 2) was located on the long (*q*) arm of chromosome 15 and was the second visible band (2) in the first region (1) from the centromere. Ideograms showing all of the about 400 visible chromosomal bands were included in the conference report.[15]

Importantly, in addition to making each human chromosome distinguishable under the microscope, cytogeneticists also used this standardized banding nomenclature as a set of landmarks for human gene mapping. The physical landscape of each chromosomal continent was carefully observed under the microscope and standardized at the 1971 Paris conference. While the number of visible bands tripled over the next decade, the basic nomenclature system agreed upon in 1971 remained in place, and in the current decade it continues to play a significant role in shaping geographic conceptions of the human chromosomes and genome. These ideograms provided medical geneticists with common coordinates and were the basis for the construction of a chromosomal infrastructure, within which all human genes and disorders were mapped.

Mapping Genetic Traits

Genetics is the study of variation and its inheritance. During the first half of the twentieth century, the examination of genetic variants was limited to phenotypes. With the rise of human cytogenetics, the search for variation was extended to the microscopic observation of chromosomes. At first, the mapping of traits to specific chromosomes occurred in fits and starts and resulted from fortuitous findings during cytogenetic analysis. Roger Donahue, a doctoral student with Victor McKusick, mapped the first gene to a human chromosome (other than X) in 1968. As part of his cytogenetics training, Donahue was examining his own chromosomes under the microscope when he identified a visible variant: one copy of chromosome 1 was shorter than the other. Next he examined the chromosomes of his family, and found the same variation in some members. He noted that the chromosomal marker appeared only in family members with a phenotypic variation involving the Duffy blood group. The chromosomal marker that Donahue had identified was limited to his family, but the Duffy blood group existed in all humans. Thus, other traits inherited with the Duffy blood group variants were also located on chromosome 1.[16]

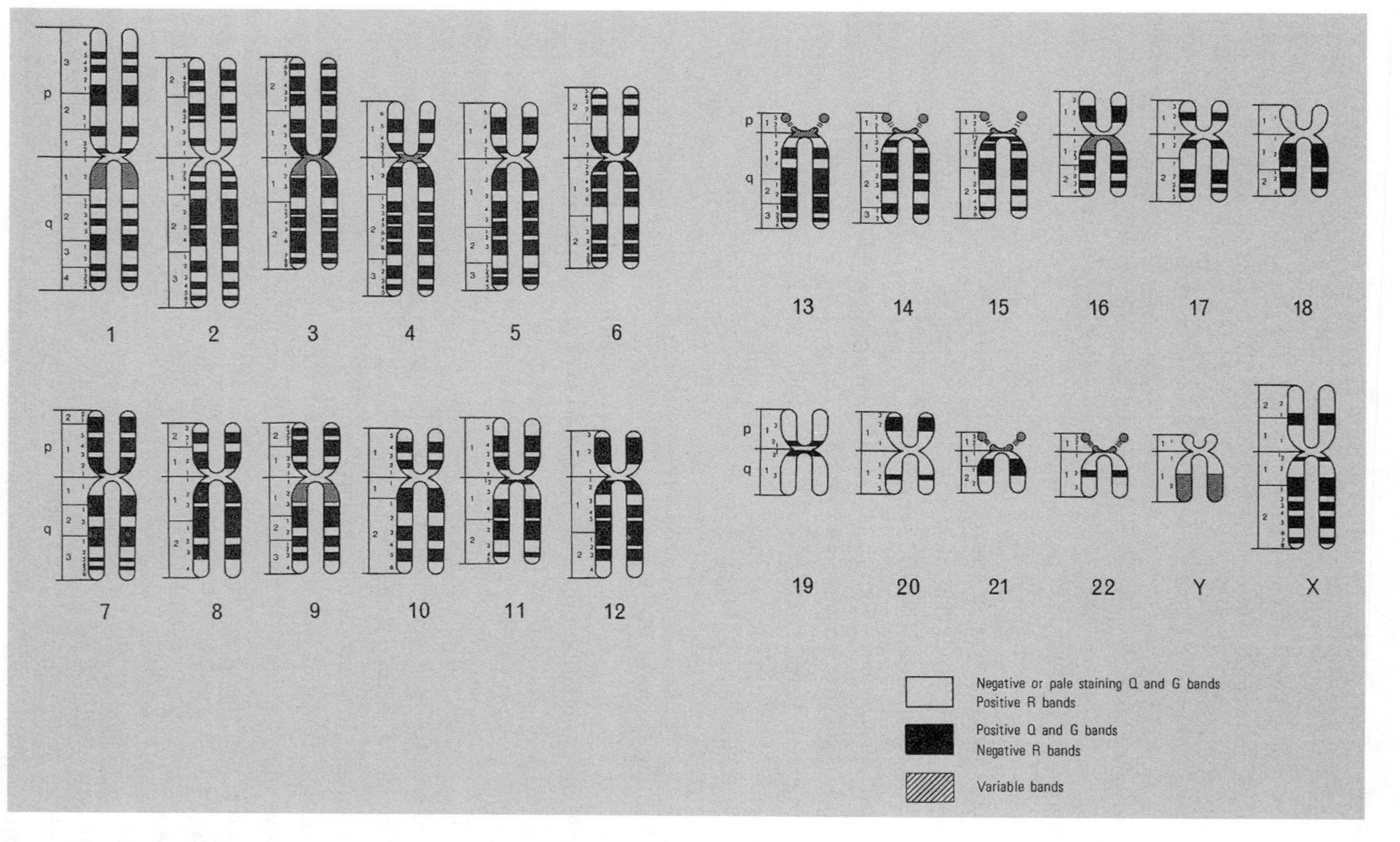

Set of banded chromosomal ideograms from the 1971 Paris Conference on Standardization in Human Cytogenetics. *Source:* Paris Conference (1971), "Standardization in Human Cytogenetics," *Birth Defects Original Article Series* 8, no. 7 (1972): 18–19. Reprinted with permission from March of Dimes Foundation.

The process of mapping genes by linkage had been ongoing throughout the twentieth century. Geneticist Thomas Hunt Morgan conducted an early and influential genetic mapping project in *Drosophila* flies during the second and third decades of the century. Morgan believed that when genetic traits were repeatedly inherited together, it implied that they must be physically linked. Over the coming decades, four linkage groups were identified and characterized in *Drosophila*. The linkage groups corresponded to four chromosomes. In 1933 cytogeneticist T. S. Painter reported on a technique for visualizing the giant polytene chromosomes of *Drosophila*, which comprised more than 100 copies of each chromosome bound together in one structure. When stained, these polytene chromosomes exhibited banding patterns, which were used as landmarks for mapping genetic traits. Similar banding patterns were not elicited in human chromosomes until the 1970s.[17]

Before Donahue's family studies involving chromosome 1 in the late 1960s, only one genetic linkage group had been established in humans, that of the X chromosome. Females possess two X chromosomes, whereas males have just one. Thus, while recessive traits on the X chromosome are often masked in females, they are almost always expressed phenotypically in males. Medical geneticists were thus able to infer that traits such as color blindness, hemophilia, and some forms of intellectual disability, all of which primarily affected males, were X-linked. By this approach, several X-linked genes were identified during the twentieth century. Like all human chromosomes, the X was thought to occasionally be involved in rearrangements, in which pieces of genetic material were swapped within chromosome pairs, and sometimes with different chromosomes (e.g., between 9 and 14). Geneticists believed that when genes were farther apart along a chromosome, they were more likely to be rearranged in this process. Based on how frequently rearrangements occurred among traits, human geneticists attempted to determine the approximate relative location of certain genes along the X chromosome. Similarly, in the early twentieth century, geneticists in Morgan's laboratory had used this approach to map fly genes.

Fly geneticists had many advantages over human geneticists. Flies had a very short reproductive cycle and could be selectively bred. In addition, during the late 1920s, geneticist Hermann J. Muller had demonstrated that X-rays could be used to increase the mutation and recombination rates in flies. Unable to actively interfere with reproduction or irradiate their subjects, human geneticists were left having to scan the chromosomes of individuals to look for interesting variants and rearrangements. What human geneticists required, in order to speed up this process, was a way of facilitating genetic recombination without having to deal with human reproduction. They needed, as J. B. S. Haldane termed it, an "alternative to sex."[18]

Somatic Cell Hybridization

In 1956 human cytogeneticists demonstrated that during sexual reproduction two genetically distinct gametes (one egg and one sperm), with 23 chromosomes each, fused to form one with 46. In the process, genetic traits from the two parents were recombined. During the 1960s innovations in human somatic (nonsex) cell genetics offered opportunities to recreate genetic recombination outside of the body, without using reproductive cells. These innovations began with rodents. Mouse geneticists noticed around 1960 that when two cells lines were grown together in the same flask, they occasionally fused, resulting in the combination of the cells' individual genetic traits into a hybrid cell line, which continued to reproduce. To isolate these hybrids, mouse geneticists grew two different cell lines in media that lacked two distinct nutrients needed for survival. Each cell line lacked the gene to make one of these essential nutrients. Rarely, hybrids formed between the cell lines, and the hybrid cells thrived because, in combining their genomes, they came to possess both of the genes they needed to survive.[19]

The hybridization of somatic cells was not limited to individual rodents. As historian Hannah Landecker described, biologists in the 1960s found that there was little barrier between different species at the cellular level. While human and mouse organs were quite distinct, the cells from these two species had no trouble hybridizing and exchanging genetic material. At the cellular level, biological components were surprisingly malleable and interchangeable. As a result, human-rodent hybrid cells were adopted as important technologies for gene mapping.[20]

In 1965 Oxford cytogeneticists Henry Harris and J. F. Watkins reported that somatic cell hybridization could be enhanced by the introduction of a virus into cell culture. Later that year, Harris and Watkins, working with their colleague Charles Ford, demonstrated that hybrids combining human and rodent cell lines could also be created in vitro using a viral catalyst. When researchers allowed these hybrids to divide in cell culture, they did not retain a full complement of both human and rodent chromosomes for very long. Rather, most human chromosomes were lost over the generations, while the rodent genome was retained in full, keeping the cell alive.[21]

Cytogeneticists quickly realized the potential value of this unusual situation. In effect, these hybrid cells brought about the random isolation of specific human chromosomes. For instance, if one of these cell lines contained a full set of mouse chromosomes but only human chromosome 7, cytogeneticists could assume that any human protein identified in cell culture was produced by a gene on chromosome 7. Also, if cells required a certain human gene to survive in a specific medium, then only hybrid cells retaining the human chromosome where that gene was located would survive. New York University researchers Mary Weiss and Howard Green concluded in a 1967 paper on human-rodent cell hybrids that the "study of clones containing a small number of human chromosomes should permit the localization of other human genes."[22]

A major roadblock in the late 1960s to using somatic cell hybrids for mapping human genes was that most chromosomes remained indistinguishable under the microscope. In 1967 Weiss and Green used somatic cell hybridization to identify the chromosomal location of the human enzyme thymidine kinase, which is involved in DNA synthesis. In their experimental setup, thymidine kinase was necessary for cell growth in some cell media but lethal in others. Weiss and Green concluded that the human thymidine kinase gene was located among the C group of chromosomes, 6–12 and X. The next year, Johns Hopkins researchers Barbara Ruben Migeon and Carol S. Miller came to a different result-based hybrid analysis. They argued that the human thymidine kinase gene was on one of the E group chromosomes, 17 and 18.[23] Somatic cell hybridization clearly offered a powerful tool for isolating the human chromosomes on which certain genes were located. However, if the chromosomes could not be individually recognized under the microscope, then uncertainty about which "continent" a gene was on would remain.[24]

The introduction of chromosomal banding during the early 1970s, and the subsequent development of a chromosomal nomenclature system, made it possible for somatic cell hybridization to concretely contribute to human gene mapping. With the aid of banding, the gene for thymidine kinase was mapped to chromosome 17 and the gene for lactate dehydrogenase, a metabolic enzyme, to chromosome 11.[25] Neither of these findings had any immediate clinical relevance. However, each association of a human enzyme gene with a certain chromosome enhanced the mapping system, allowing for the selection of hybrids with certain chromosomes. Soon panels of somatic cell hybrids containing different combinations of human chromosomes were developed and were used to determine the chromosomes on which human genes were located.[26]

Mapping genes to specific human chromosomes was just the first step. Human cytogeneticists next hoped to map genes to specific regions within each chromosome, defined by the 1971 Paris banding nomenclature. To accomplish this, researchers sought to develop somatic cell hybrids containing specific human chromosome fragments. If a gene was known to be on chromosome 4, and its expression was detected when only one small portion of that chromosome was present in a specific cell hybrid, then that gene could be assumed to be located in that chromosomal region. Chromosome fragments formed in translocation events in which parts of two different chromosomes became attached. Cytogeneticists knew that such rearrangements occurred infrequently and were not likely to happen randomly in somatic cell hybrid lines. Instead, they relied on human cell lines, often from the clinical setting, to provide specific rearrangements that could then be introduced into somatic cell hybrid lines. Such rearrangements occasionally came about in human cells and were sometimes scanned for in the clinical context. Many were benign, but others caused disorders or repeated miscarriages.[27]

During the 1970s the collection of a wide variety of human-rodent somatic cell hybrid lines played a central role in facilitating advances in human gene mapping. Cytogeneticists built libraries of these hybrids, which contained varying complements of human chromosomes. Also important was the development of a repository of human mutant cell lines funded by the National Institutes of Health. In 1972 virologist Lewis L. Coriell founded the Human Genetic Mutant Cell Repository at his Institute for Medical Research in New Jersey. The repository was initially intended to collect cell lines from individuals with genetic disorders, but its focus quickly turned to aiding in the collection of cells containing chromosomal rearrangements that could be useful for somatic cell hybridization and gene mapping. Short descriptions of the repository in various biomedical journals encouraged researchers to submit samples of cells with genetic mutations for storage in liquid nitrogen at the institute. Upon request, these samples would be held for up to a year before being made available to the broader research community.[28]

Throughout the 1970s the repository regularly published short notes that were part advertisement and part research report. Featured in the journal *Cytogenetics and Cell Genetics,* these short write-ups provided descriptions of the repository samples that were available (for a $20 processing fee, plus shipping). Each note also counted as a publication for the researchers who initially submitted the samples, which was an important consideration within the awards system of academic medical genetics. The short descriptions offered information about the clinical features of the patient from whom the sample was acquired, an explanation of the chromosomal abnormality present in the cells, and suggestions about how researchers might use the particular sample, for instance, for gene mapping studies of chromosomes 1 and 17.[29]

The repository also published an annual catalog of its cell lines. In addition to a table of available samples, it included an explanation of how researchers could go about submitting new cell lines. As part of the process, they filled out a standardized form describing the patient and the tissue from which the cells originated, as well as various clinical features and a detailed description of any known genetic abnormalities. The permission form also asked submitters to confirm that "appropriate consent was obtained from the patient from whom the cells were originally obtained, or [could] be reasonably inferred, for use of these cells for diagnosis, research, teaching, or therapy."[30] The nature of this consent and how it was obtained was left to the discretion of the submitting researchers and their institution. With no formal consenting process in place, one can assume that patients' awareness of what they were contributing to was limited. Even with more stringent protections, however, it would have been difficult for clinical providers to predict the uses, beyond those relating to a specific disorder, to which any cell line would eventually be put.

During the 1970s and 1980s, the Human Genetic Mutant Cell Repository played a central role in collecting cell lines containing chromosomal fragments that had been acquired for clinical study but which were then put to use in the laboratories of human gene mappers. Over this time, the repository shifted from collecting cytogenetic abnormalities in order to facilitate research on specific pathologies, to collecting cell lines that could aid cytogeneticists in studying the normal makeup of the human genome. Indeed, whether individual patients realized it or not, their cell lines, which had been donated in the context of clinical diagnosis, often become the raw material of basic human cytogenetics research. Throughout the postwar period, the translation of knowledge from basic genetics research into clinical practice was a well-funded and publicized pursuit of biomedicine. The Human Genetic Mutation Cell Repository and its contributions of clinical materials to basic research offer an important example of how this translation of findings and resources occurred in both directions. Indeed, as many scholars have demonstrated, biomedical knowledge production has long been a two-way street linking the laboratory and the clinic.[31]

Collaborative Gene Mapping

During the early 1970s, a German-trained physician named Uta Francke was uniquely well positioned to contribute to the burgeoning field of human gene mapping. As a genetics fellow at the University of California, Los Angeles, Francke used Q banding to characterize chromosomal translocations from clinical patients. Soon thereafter, Francke applied her growing expertise with chromosomal banding to the characterization of mouse chromosomes as well. In 1971, after moving to San Diego, where her husband had a position at the Salk Institute, Francke sought to enter a medical residency program but could not because her medical training had taken place in Germany and she was not licensed in California. Instead, she took a job in the laboratory of William Nyhan at the University of California, San Diego (UCSD), where she performed biochemical carrier screening for the X-linked disorder Lesch-Nyhan syndrome, which her supervisor had helped to identify.[32]

It was at UCSD that Francke became aware of somatic cell hybridization from her colleague Jerry Schneider, who had recently returned from a sabbatical in Paris with Boris Ephrussi, one of its innovators. Given her experience with human and mouse chromosome banding, access to translocations from clinical patients, and knowledge of biochemical genetic markers, Francke had exactly the expertise needed to regionally map human genes. She received a research grant from the National Institutes of Health (NIH) for a gene mapping project. Francke also became an assistant professor in pediatrics at UCSD. The NIH review board was so enthusiastic about her proposal that they recommended that her budget be increased so she could purchase the best possible microscope. In an era of shifting US government support for basic biological research, human gene mapping appeared to be on solid footing.[33]

As Francke's work got under way, she was invited to present at an international workshop on human gene mapping, which was holding its second annual meeting in 1974. Frank Ruddle and his colleagues at Yale University had hosted the First International Workshop on Human Gene Mapping in June 1973. With the help of Victor McKusick, the conference received funding from the March of Dimes, which had moved into funding research on birth defects in the 1960s, following the eradication of polio in the United States. Conference proceedings were printed in a special issue of the journal *Cytogenetics and Cell Genetics*. The workshop featured presentations of unpublished human gene mapping research, with a focus on the localization of genes to specific chromosomes and the identification of gene linkage groups. In addition, committees were convened to keep track of international progress in gene mapping. At the 1973 conference, individual committees were formed to report on chromosomes 1 and X, which had seen the most mapping progress to that point. Much of the human gene mapping done at this time relied on somatic cell hybridization, which was further supplemented by more traditional family linkage studies. By the time of the first meeting, the yield was already quite high, with 28 provisional and 31 confirmed human genes mapped to specific chromosomes other than X.[34]

The first three human gene mapping workshops were held annually, with the second meeting in Rotterdam and the third in Baltimore, hosted by McKusick. Researchers and funding for human gene mapping came from all over Europe and North America. The 1973 workshop featured more than 60 participants, who had received support for their gene mapping work from the National Science Foundation, the Atomic Energy Commission, the March of Dimes, and the NIH in the United States, the Wellcome Trust in the United Kingdom, the Medical Research Council of Canada, and the Netherlands Organization for Fundamental Medical Research. At the 1974 meeting, additional research funding for participants came from the American Cancer Society, as well as government funding organizations in Norway, Sweden, France, Germany, and Scotland. Human gene mapping drew interest from a variety of funding institutions, which clearly saw this pursuit, and its potential for producing biological and medical breakthroughs, as falling into their purview.[35] By the third meeting, participation had grown to more than 100, leading to the decision to decrease the number of paper presentations at future meetings, in favor of poster presentations and a more targeted focus on committee work relating to progress on specific chromosomes.[36]

Uta Francke was an active participant at the human gene mapping workshops. Her use of somatic cell hybridization and translocated human chromosome fragments contributed to the regional mapping of many genes. Francke was also a member of multiple committees at the workshops, which tracked the mapping of various chromosomes. Upon receiving the American Society for Human Genetics most prestigious Allan Award in 2012, Francke noted that in the decades before the Human Genome Project, the human gene mapping workshops provided an extensive physical reference map of the human genome for future use. Between 1973 and 1975, the number of mapped human genes doubled, and it doubled again by the end of the decade. Over this time, somatic cell hybridization was the primary means of locating genes on chromosomes.[37]

The Thrill of Mapping

In 1976 an important milestone for the human gene mapping community was reached: at least one gene had been mapped to every human chromosome. McKusick and Ruddle celebrated this accomplishment in *Science* the next year. Their article included a gene map that used banded chromosomal ideograms to depict the location of each mapped human gene.[38] These ideograms have always been fundamental to the practices of human gene mapping. As Francke put it, "One reason for generating chromosome-banding maps is to have a template onto which genes can be placed.[39] McKusick had begun making these maps a few years earlier and had included one in the 1975 Human Gene Mapping Workshop report. In constructing his gene maps, he took advantage of the infrastructure provided by the 1971 Paris banding nomenclature. McKusick's human gene maps did more than just tabulate on which chromosome each gene had been found; they also provided a visual approximation of where along a chromosome each gene was located. This representation captured the cartographic orientation of human gene mapping, demonstrating what had been accomplished and what areas remained to be explored. By 1977 he was able to show that at least one flag had been planted on every chromosome continent.

In the mid-1970s, the human gene mapping community was small, their techniques were complex, and most of their findings were obscure. However, cartographic metaphors helped to make the accomplishments and goals of gene mapping accessible to journalistic coverage. Harold M. Schmeck Jr, a science reporter for the *New York Times,* drew upon McKusick and Ruddle's language in his coverage of human gene mapping. In 1975 he wrote, "The world's oceans have been charted and its continents mapped. Today, a new kind of cartography is in progress, the mapping of the human genes and chromosomes." Schmeck highlighted the importance of chromosome banding, noting, "Sophisticated methods of staining or otherwise marking chromosomes brings out the unique structure of each. Study of the structures allows geneticists and chemists not only to find the specific chromosome, but to locate on it where the gene for a given chemical trait is to be found. In the past several years, such studies have contributed to a growing gene map that now incorporates about 100 specific sites, with one or more on almost every chromosome."[40] Schmeck's articles brought human gene mapping and the infrastructure upon which it depended to broader public attention during the mid-1970s. With his columns, Schmeck aided human gene mappers by highlighting the potential long-term social and scientific value of their project.

McKusick and Ruddle sought to broaden the appeal of human gene mapping within the scientific community as well by characterizing its importance as an intellectual and basic research pursuit. In their 1977 *Science* paper, they noted, "Mapping the human genome, like any uncharted terrain, is a challenge to the human intellect. Inevitably, information on the detailed gene anatomy of the chromosomes will also have usefulness. Practical value of the information need not and should not be a primary concern."[41] Gene mapping was presented as a cognitively rewarding scientific project. McKusick considered it "a matter of considerable intellectual satisfaction that we know that the gene for Rh blood type is on the short arm of chromosome 1, that the gene for ABO blood type is on the end of the long arm of chromosome 9." Identifying the genetic inhabitants of previously unexplored "chromosomal continents" drove their interest.[42]

When addressing scientific colleagues, Ruddle in particular highlighted the thrill of gene mapping as a new form of knowledge production, over

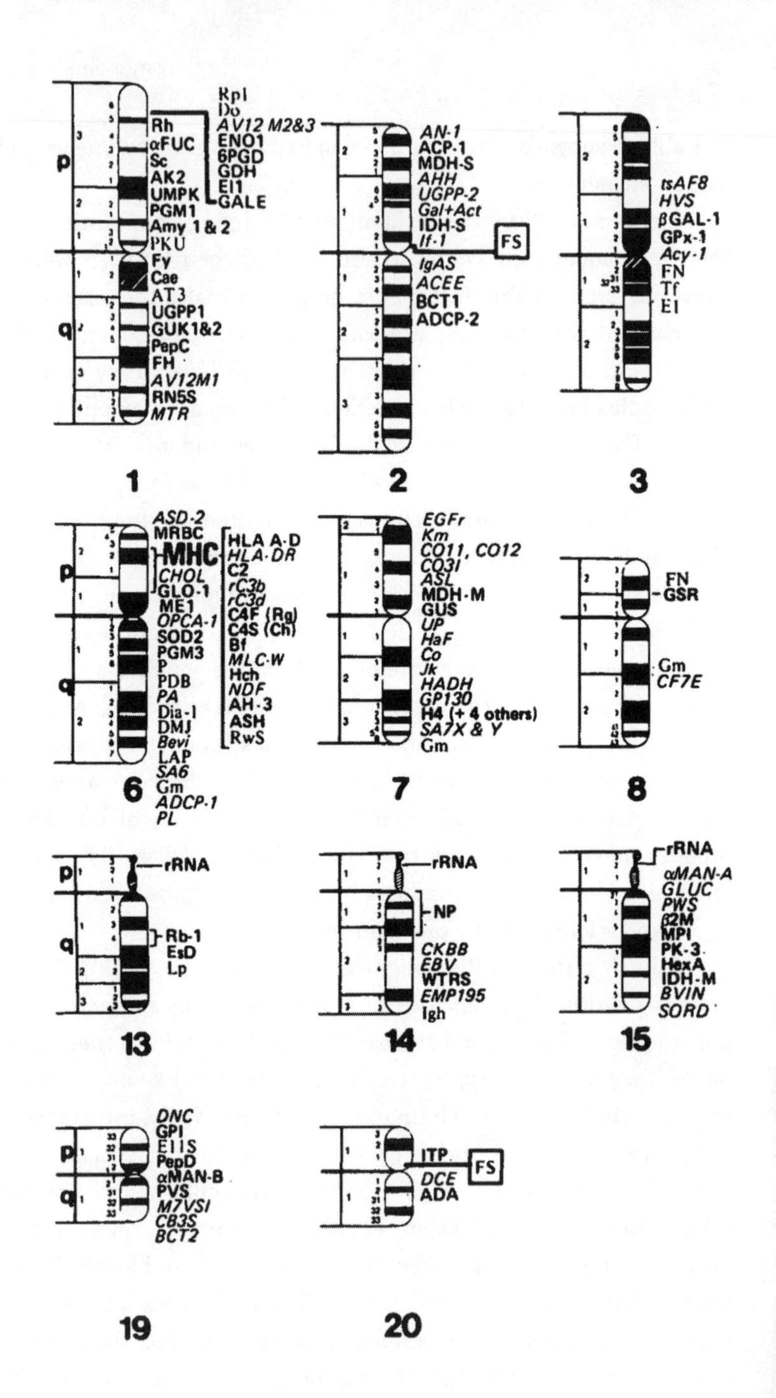

p
q
Rpl
Do
AV12 M2&3
ENO1
6PGD
GDH
El1
GALE
Rh
αFUC
Sc
AK2
UMPK
PGM1
Amy 1 & 2
PKU
Fy
Cae
AT3
UGPP1
GUK1&2
PepC
FH
AV12M1
RN5S
MTR
1
AN-1
ACP-1
MDH-S
AHH
UGPP-2
Gal+Act
IDH-S
If-1
FS
IgAS
ACEE
BCT1
ADCP-2
2
tsAF8
HVS
βGAL-1
GPx-1
Acy-1
FN
Tf
El
3
ASD-2
MRBC
MHC
CHOL
GLO-1
ME1
OPCA-1
SOD2
PGM3
P
PDB
PA
Dia-1
DMJ
Bevi
LAP
SA6
Gm
ADCP-1
PL
HLA A-D
HLA-DR
C2
rC3b
rC3d
C4F (Rg)
C4S (Ch)
Bf
MLC-W
Hch
NDF
AH-3
ASH
RwS
6
EGFr
Km
CO11, CO12
CO3I
ASL
MDH-M
GUS
UP
HaF
Co
Jk
HADH
GP130
H4 (+ 4 others)
SA7X & Y
Gm
7
FN
GSR
Gm
CF7E
8
rRNA
Rb-1
EsD
Lp
13
rRNA
NP
CKBB
EBV
WTRS
EMP195
Igh
14
rRNA
αMAN-A
GLUC
PWS
β2M
MPI
PK-3
HexA
IDH-M
BVIN
SORD
15
DNC
GPI
E11S
PepD
αMAN-B
PVS
M7VSI
CB3S
BCT2
19
ITP
FS
DCE
ADA
20

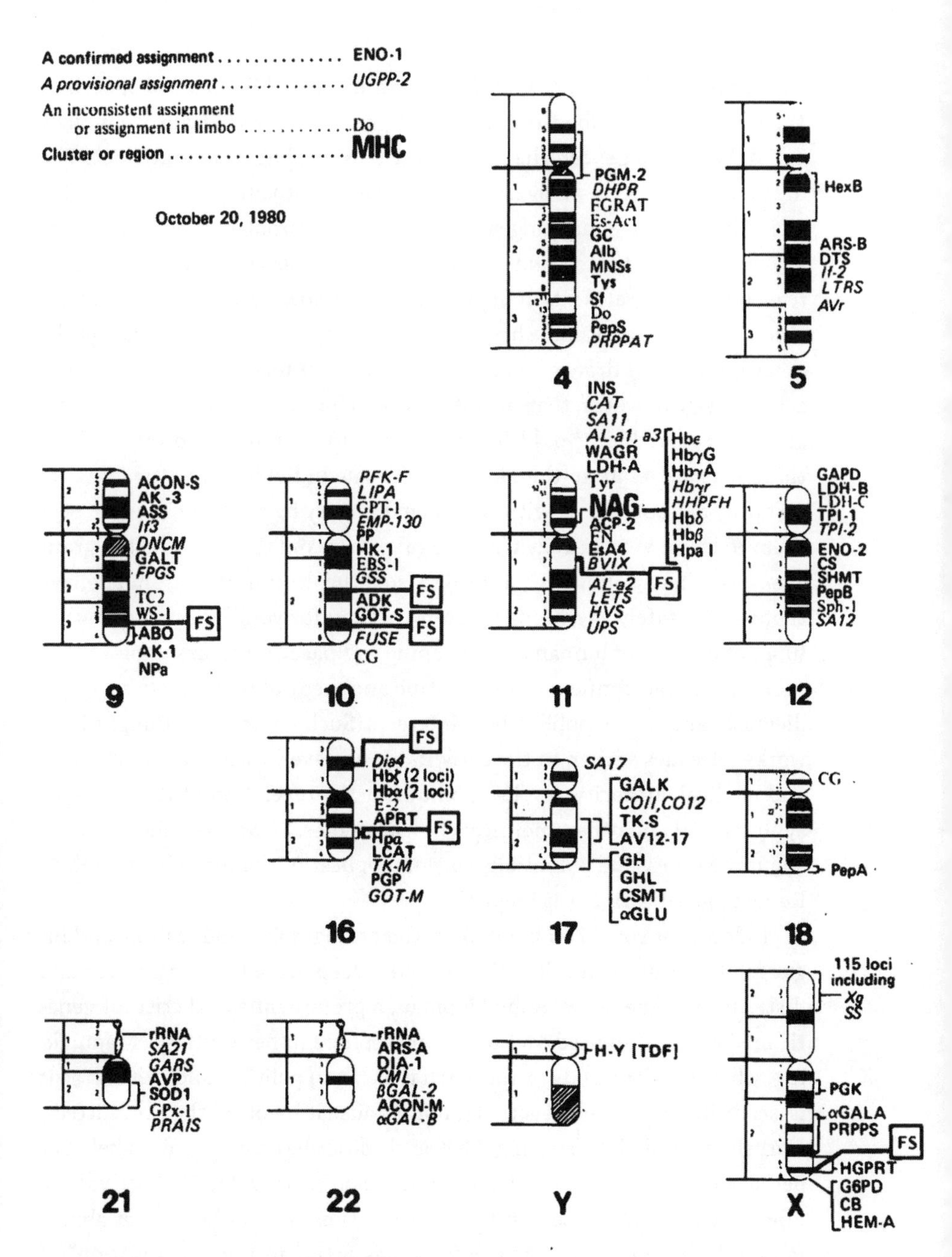

Human gene map built on banded chromosomal ideograms by Victor McKusick in 1980. *Source:* V. A. McKusick, "The Anatomy of the Human Genome," *Journal of Heredity* 71, no. 6 (1980): 376. Reprinted with permission from Oxford University Press.

any concerns about its practical value. In his own Allan Award speech in 1983, Ruddle described the appeal of gene mapping: "I am frequently asked, Why map genes? There are many reasons, but three stand out. The first relates to the challenge. Many of us were motivated to develop methods of human gene mapping simply because the problem seemed insuperable at the outset. A second motivation is esthetic satisfaction. For some reason we gain gratification in placing a well-known gene in its correct position on the map." Ruddle began his answer by once again expressing the great intellectual draw of mapping an unknown territory. He went on to acknowledge however, that the interests of a few human gene mappers and genetics researchers would be insufficient to sustain the project: "Both these reasons relate to the motivation of the individual investigator, and certainly are not shared with the same intensity by all. Nor can we expect them to count very heavily with the taxpayers and their agents: the grant agencies. There is, however, a third reason: the utilitarian reason. I believe that we can safely say, 'Gene mapping is good for you!' "[43] Ruddle's rather flippant defense of human gene mapping was based on an assumption that there was broad confidence in scientific and medical researchers to act judiciously and in the public's best interests. Such an argument might have worked decades earlier. In the early 1980s however, with trust in scientific and medical authority at a low point after the revelation of the Tuskegee syphilis trial and other unethical clinical studies, as well as public concern about the safety and potentially eugenic applications of recombinant DNA, Ruddle's justification was inept.[44]

Indeed, for some commentators, the cartographic and utilitarian language of McKusick and Ruddle aroused a deep-seated concern about and distrust in biomedicine. Abby Lippman, a prominent social critic of genetic medicine, referenced McKusick's metaphors in her writing: "Mapmaking, whether of the body or the earth is as much political and cultural as it is 'scientific'. It is a social activity, not because it involved the committees, workshops and data sharing McKusick describes under this label, but because it is an expression of and influence on social values." Those values, Lippman argued, focused on the identification and eradication of abnormality. In her view, genetic medicine was active in the "colonization" of health and illness, which "gives mapmakers, their funders and the purveyors of their products tremendous power (and wealth) for defining how we think of ourselves and others." The metaphor of mapping, as Lippman saw it, was very effective in legitimating the reductive and deterministic assumptions of medical genetics, which contributed to eugenic efforts to cut the social costs associated with disease and "establish control over our genetic endowment."[45] Lippman was troubled both by the insularity of the human gene mapping community and by how successful they had been in gaining financial support and shaping conversations about disease. Indeed, the one mutation–one disorder narrative, as she readily acknowledged, was quite powerful in late-twentieth-century biomedicine.

Anatomizing the Genome

Recognizing the limited appeal of mapping metaphors, McKusick took the conversation in a different direction, replacing cartographic allusions with anatomical ones. This framing of human gene mapping and its value aimed to broaden the tent of who was interested in becoming involved in its efforts and funding. As McKusick put it in a 1979 American Society of Human Genetics lecture, "Mapping previously uncharted terrain holds a certain fascination for many. . . . However, I have substituted an equally apt simile. The gene map of the man's chromosome is part of his anatomy." McKusick went on to note that the human genome was anatomical in multiple senses: "There is a morbid anatomy, a comparative anatomy, a functional anatomy, a developmental anatomy, and even, if not a surgical anatomy, at least the beginnings of an applied anatomy."[46] McKusick thus presented gene mapping as having value for researchers in multiple areas. Evolutionary biologists would consider the arrangement of genes along the chromosomes of humans and other primates as they studied comparative anatomy, while physiologists and developmental biologists would examine how the ordering of genes was related to their functionality in bodily systems at various stages of development. He argued that gene mapping was a deserving pursuit. It amounted to much more than mere "stamp collecting," a common critique of data collection for its own sake.[47]

McKusick also drew on anatomical references to place gene mapping within the larger scope of medical tradition and history, making it more broadly accessible to physicians. He proclaimed in 1981, "Since the time of Vesalius anatomists and physiologists have been charting the internal features of the human body with ever increasing detail. . . . We have now arrived at a sort of 'last frontier' of anatomy, that of the human genome."[48] This passage referred to the sixteenth-century physician Andreas Vesalius, who was known for his public dissections and his compendium of updated human anatomical figures. McKusick used the frontispiece of Vesalius's multivolume text *On the Fabric of the Human Body* in his paper "The Human Genome through the Eyes of Mercator and Vesalius." The image depicted Vesalius performing an anatomical demonstration in a public forum by pointing to a newly dissected human body. This was in stark contrast to the long-standing practice of anatomical instructors, who traditionally taught by reading from the 1,400-year-old text of Galen while standing far away from the dissected body.[49]

Vesalius was regarded in the Western medical canon as a revolutionary figure who brought direct observation of the dissected human body back to the forefront of anatomical research and teaching. Historian Katherine Park, in her book *Secrets of Women,* provided a revisionist perspective on Renaissance era dissection, counting Vesalius among sixteenth-century figures who have long been misrepresented as heroes who "braved persecution and censure in the service of art and science." Continued reference to these individuals, Park argued, did, "important cultural work," providing "foundation stories that confirm deep-seated Western institutions about the scientific origins of modernity—institutions that continue to inform the writing of even specialists in the field."[50] Indeed, postwar physicians regarded Vesalius as a revolutionary figure who had a significant impact on centuries of medical thinking and practice.

McKusick was not alone in his Vesalian references to human gene mapping. Medical geneticist Charles Scriver drew on similar language in his writing and advocacy during the 1980s, ascribing a "neo-Vesalian" character to human gene mapping.[51] He found this framing to be rhetorically valuable when making the pitch to potential funders of human gene mapping.[52] Just as the work of Vesalius was understood as having reshaped medicine in the sixteenth century and beyond, McKusick and Scriver presented anatomical exploration of the human genome as likely to have revolutionary implications for medicine in the decades to come.

McKusick's references to the many anatomies of the human genome were also reflective of his larger project to bridge long-standing divides in human genetics between physicians and biologists. In his 1975 presidential address to the American Society of Human Genetics, he lamented, "[C. P.] Snow would say that we have had two cultures within the American Society of Human Genetics: the non-M.D. and the M.D. . . . In the 1970s, let us hope we are achieving a state of mutual respect and intimate collaboration between the two cultures." McKusick pointed to medical genetics assessment as a valuable forum for bringing together the two cultures: "Two related questions are always in mind: How can we get at the underlying defect in this patient's disorder? and What of general applicability does this patient have to teach us?"[53] Indeed, just as chromosomal rearrangements from patients became the raw material of gene mapping research, clinical cases could point basic human geneticists toward new genetic entities and questions.

More broadly, McKusick understood that maps often played a central role in bringing people together. For a city, a nation, or a professional community, maps reflected a sense of what defined and bound together a group of people. In science, maps provided a picture of what had been accomplished and what remained to explore. A partially filled in map directly conveyed a need for completion and was used as a strong argument for continued funding. McKusick's chromosome-level maps showed gene mappers and their funders what was left to be colonized.[54] As a member of the Howard Hughes Medical Institute Advisory Board in the 1980s, Scriver played an important role in encouraging continued investment in human gene mapping. He later explained, "There was lots of initial episodic work, where a certain gene might be mapped to a particular region of a chromosome, and so a mosaic was being built up. I was interested in seeing the whole picture being completed."[55] During the 1970s and early 1980s, Ruddle, McKusick, and others sought to add momentum to the ongoing process of human gene mapping. In doing so, they pointed to the potential for human gene mapping to achieve a variety of scientific aims, while promising that it would also improve the clinical treatment and prevention of disease.

Preventing Genetic Disease

While much was made of the potential clinical value of human gene mapping in combating inborn disorders, during the 1970s few approaches for preventing birth defects were actually based on genetic testing. Testing that did involve the analysis of genetic material was primarily focused on identifying large or distinctive chromosomal abnormalities under the microscope, including trisomy 21 and the fragile X site. Many medical geneticists instead sought to identify biochemical markers for genetic disorders such as Tay-Sachs disease and phenylketonuria (PKU), to facilitate their prevention. PKU was one of many Mendelian recessive genetic disorders caused by the inheritance of two mutant copies of a particular gene. Medical geneticists had known for decades that this gene produced an enzyme that was necessary for the breakdown of the amino acid phenylalanine. Without the enzyme, phenylalanine, which was present in many foods, built up quickly in the body and caused severe mental retardation. The impacts of PKU could be prevented only if a low-phenylalanine diet was instituted from birth. For many years after this discovery, the production of such a diet was considered to be economically infeasible.[56]

In 1960 microbiologist Robert Guthrie developed a biochemical blood test for the early infant diagnosis of PKU. The test did not identify mutant genes but rather the first signs of phenylalanine buildup in newborns. By this time, a feasible low-phenylalanine diet had been developed, and Guthrie's PKU blood test soon began to be mandated for newborn screening in most US states. PKU thus became a rare example of an inherited genetic disorder that could largely be prevented when detected after birth. The necessary diet made for a difficult lifestyle, but nonetheless it became an exemplar of how genetic and biochemical knowledge could be used to prevent mental retardation. PKU was one of many disorders that medical geneticists targeted during the mid to late twentieth century to demonstrate the clinical value of their research. Individually, these diseases were rare, but if many others were identified and prevented in the same way, medical geneticists argued, the impacts on public health would be significant.[57]

For most genetic diseases during the 1970s, prevention required some combination of carrier screening, prenatal testing, and abortion. In 1969 researchers determined that Tay-Sachs disease, another Mendelian recessive disorder, which was lethal in early childhood, was caused by the absence of the enzyme Hex-A. Clinical trials showed that administering this enzyme to affected individuals, no matter how early in development, could not stop Tay-Sachs disease. Blood screening for lower-than-normal levels of Hex-A however, could be used to identify carriers of the recessive genetic trait. To prevent Tay-Sachs disease, carriers could abstain from reproduction or undergo amniocentesis to test for the fetal presence of Hex-A prenatally. If the enzyme was absent from the fetus, the only preventive measure was abortion. Faced with this difficult choice, Jewish groups, because Jews were at greatly increased risk for Tay-Sachs disease, organized carrier-screening programs, in some cases to prevent carriers from marrying.[58]

A major complication in preventing Mendelian recessive disorders, when biochemical tests for them existed, was identifying carriers before an affected child was born. Outside of disorders that were common in certain ethnic communities, most couples had no reason to believe that they shared a disease-causing recessive mutation; they found this out only when they had an affected child. In a 1976 *New England Journal of Medicine* review article, clinical geneticist Aubrey Milunsky identified 25 metabolic disorders that had been successfully diagnosed prenatally and another 55 that could possibly be detected in the future. These conditions, however, represented just a small fraction of Mendelian recessive single-gene disorders. Milunsky admitted in 1981, "Although almost 3,000 monogenetic disorders have been recognized, less than 5 percent are currently diagnosable prenatally."[59] Among these were severe and relatively common disorders such as cystic fibrosis, which is a severe respiratory disorder that often leads to early death, and Huntington's disease, a neurodegenerative condition that strikes in middle age. Preventing these single-gene disorders first required locating their genetic cause in the genome.

As described in the life history of fragile X syndrome (chapter 1), chromosome studies occasionally offered clues about the location of disease-causing genes. In the 1960s medical geneticists identified multiple forms of cancer that were associated with chromosomal abnormalities. Included was the Philadelphia chromosome, first identified in patients with chronic myeloid leukemia by University of Pennsylvania researchers Peter Nowell and David Hungerford. After banding was introduced, University of Chicago cytogeneticist Janet Rowley demonstrated that this unusual chromosome resulted from a translocation of chromosomes 9 and 22, suggesting that a causative gene might be located where these two chromosomes crossed. Another translocation similarly suggested the location of the cancer gene that causes Burkitt lymphoma, along chromosomes 8 and 14. In each instance, chromosomal analysis under the microscope pointed to a set of chromosomal bands where a disease-causing gene was likely to be located.[60]

During the 1970s many additional human chromosomal abnormalities were identified, but few were seen regularly enough to be associated with a reoccurring clinical disorder. A major aim of human gene mapping at this time was to associate genetic diseases with increasingly targeted portions of the human genome, such that one day causative genes could be identified. As Harold Schmeck reported in the *New York Times,* "genetic mapping is expected to have important future uses in promoting understanding of human development, diagnosing birth defects early in fetal life and predicting a person's inherent susceptibility to diseases."[61] Gene mapping was presented by its promoters, and appeared in media coverage, as an important early step in bridging the gap between what was made visible (in the clinic or under the microscope) and the molecular changes that caused disease. From here, medical geneticists pointed to a future in which, once these mutations were located in the genome, genetic engineering would be used to cure previously untreatable disorders by replacing aberrant genes with normal copies.

The Promise of Gene Surgery

During the late twentieth century, the association of a clinical condition with a particular gene or genomic location grew to be an increasingly important juncture in the life history of that disorder. The concept that many diseases had a genetic basis suggested that more universal approaches to treatment could be pursued. Physician Aubrey Milunsky, in his 1977 popular press book *Know Your Genes,* noted that gene mapping could eventually facilitate gene therapy, the introduction of normal genes in place of mutated ones. "Knowing which gene to excise and replace would depend on knowing its exact location on a particular chromosome."[62] The ability to find and replace defective genes was presented as the ultimate cure for previously hopeless genetic diseases. A short piece printed in the *New York Times* and other newspapers in 1977 similarly claimed, "Knowing the location of that many [1,000 genes], scientists believe, will contribute significantly to man's understanding of evolution and hereditary diseases. It could also facilitate 'genetic engineering,' or alterations in hereditary endowment, in human beings. Genetic engineering could be used to prevent birth defects or to treat inherited diseases."[63] In the popular press, there were many references to the future potential of gene mapping, which would be increasingly realized as more genes were successfully located.

It was at this time that a potential mechanism for genetic engineering was being developed using recombinant DNA technologies. Geneticists had begun employing restriction enzymes, which cut DNA in specific locations, in order to rearrange and even replace genes. McKusick, continuing his use of anatomical metaphors, referred to these enzymes as the "scalpel" of geneticists, an instrument that allowed researchers to dissect and examine the genome.[64] In a 1981 *Hospital Practice* article, he explained, "This truly therapeutic approach to genetic disease would involve what is sometimes called 'gene surgery' or 'gene transplantation.'" This option was no longer in the realm of science fiction, McKusick informed his readers. Rather, "when we examine the various steps that would be involved in such an 'operation,' they turn out to be either feasible by current techniques or achievable by extensions of such techniques within the foreseeable future."[65] With various techniques already at their fingertips for the transplantation of genes, what researchers needed to make gene therapy a reality was the increased knowledge and access to human genes that the ongoing mapping project was slowly making possible.

A 1982 meeting held at Cold Spring Harbor's Banbury Center brought together nearly 50 leading molecular biologists and cytogeneticists to examine "Gene Therapy—Fact and Fiction." In a book-length report on the meeting, meant for broad public consumption, physician Theodore Friedmann situated the future promise of gene therapy as the next step in the long march of medical progress: offering treatments for disease that inoculation and antibiotics could not be used to fight. Among the discussions of various technical approaches for inserting new curative genes into the human genome, and the limited array of treatments that already existed for treating genetic disorders, a presentation by Victor McKusick on gene and disease mapping was prominently featured in Friedmann's report. McKusick highlighted findings suggesting that many disorders, including cancer, which were not previously thought to be genetic, had recently been found to have a genetic basis. Paraphrasing McKusick, Friedmann noted, "Genetic disease is all around us. We no longer have the luxury of considering genetic diseases to be rare and of no consequence to most of us."[66] McKusick's message was clear: the more we look for genetic factors in human disease, the more we will find. This assumption was built into his anatomical maps of the human genome and was increasingly woven into the one-mutation-one-disorder-oriented framework and objectives of late-twentieth-century medical genetics.

Meeting participants presented human gene mapping as the first step toward realizing their promising vision of gene therapy. In Friedmann's conclusion and commentary, he stated, "Before we can develop methods for genetic manipulation or gene therapy for human disease, we have to understand the structure and function of normal and defective genes and the mechanisms for their expression. The Branbury discussions revealed that there are already techniques available which are powerful enough for us at least to imagine being able to isolate and characterize almost any gene."[67] These gene identification techniques included family linkage studies, somatic cell hybridization, and emerging molecular techniques. By the early 1980s, a decade of mapping, mostly by cytogenetic methods, had demonstrated that a completed human gene map, though still far off, was technically feasible and was a goal within reach. The infrastructure for genetic diagnosis and prevention, which had not existed 15 years earlier, was quickly being built up as "chromosomal continents" were colonized and mapped.[68]

Conclusion

In 1977 Aubrey Milunsky opened his popular press book *Know Your Genes* with a dire warning printed in capital letters, "YOU ARE A CARRIER OF FOUR TO EIGHT DIFFERENT HEREDITARY DISEASES! AND SO ARE WE ALL!" Here Milunsky was repeating a statistic that had first been popularized by geneticist Hermann J. Muller decades earlier in his 1950 presidential address to the American Society of Human Genetics. The figure made its way into the public sphere many times during the 1970s, having been repeated by medical geneticists including Kurt Hirschhorn and Fred Bergmann in interviews with the *New York Times*.[69] These clinicians were active in bringing such genetic risks to public attention because of the increasing sense that something could be done to treat or prevent many disorders if they were identified early enough.

Noting that at least 1 in 10 Americans would be affected by an inherited disease, Milunsky argued, "This does not have to be. That is why I have written this book: to elucidate the many ways in which tragedy can be prevented *in advance*. . . . Science now provides many tests that help us to prevent or, at least, successfully treat the hereditary disorder."[70] By the late 1970s, medical geneticists were able to diagnosis carriers of numerous diseases for which their offspring were known to be at high risk. Most common among these were chromosomal disorders such as Down syndrome, the incidence of which increases significantly with maternal age above 35.

Also included were certain recessive disorders, where two carrier parents knew that together they had a 25% chance of having an affected child. The vast majority of genetic disorders, however, remained completely unpredictable. Most of these conditions were rare, but as medical geneticists warned, when considered as a group, they were quite common; and some, such as cystic fibrosis and Huntington's disease, even occurred at relatively high rates.

The long-term clinical promise of human gene mapping was in facilitating the prevention of formerly unpredictable genetic diseases. Efforts to map all of the human genes during the 1970s contributed to the ongoing construction of an infrastructure that was slowly expanding the reach of genetic disease diagnosis and prevention. At the beginning of the decade, a complete human gene map still appeared to be *far from possible.* But by 1980 a complete human gene map looked to be within reach and was viewed as merely *far from finished.* With the establishment of systematic laboratory methods, the mining of valuable raw materials from the clinic held in communal repositories, and the development of various international standards and organizations, McKusick's moon-shot vision of a human gene map was becoming a reality.

Beyond completing the human gene map was the goal of sequencing every nucleotide in the genome. This vision came to fruition as the Human Genome Project (HGP) in the 1990s and similarly promised significant clinical advancements in preventing genetic disease. Scholars have noted that the HGP had its origins in molecular biology.[71] However, human cytogeneticists rightfully expressed great pride in their contributions, which stretched back to the early 1970s, to the larger human gene mapping project. Not only did human cytogeneticists create the infrastructure upon which all human genomic maps were built; they also identified many important physical landmarks along each chromosome. McKusick, who gave a paper on the existing accomplishments in human gene mapping at a 1986 symposium, later noted that his report "was an eye-opener to molecular biologists."[72] By the mid-1980s, much had already been learned about the layout of the genes along the human chromosomes from human cytogenetics work, with which molecular biologists were largely unfamiliar. Uta Francke said in her 2012 Allan Award address, "What many young people do not realize today is that when the HGP started . . . a rather dense physical map was already available to facilitate the assembly of the first human reference genome."[73] Human cytogeneticists did much during the 1970s to lay the groundwork for the Human Genome Project, both in terms of robust gene mapping and by inducing public sector interest and enthusiasm for such an undertaking.

While the discussions that eventually led to the Human Genome Project were under way during the 1980s, cytogeneticists continued to locate the causes of disorders among the human chromosomes, in some cases contributing to diagnosis and prevention. The development of higher-resolution chromosomal banding techniques in the late 1970s, discussed in chapter 3, facilitated the identification of mutations for genetic disorders that did not run in families but instead resulted from the gain or loss of very small amounts of genetic material from specific chromosomal locations. The association of multiple conditions with chromosomal abnormalities that were previously too small to be made visible was a significant event in the life histories of these disorders, bringing both greater clarity and confusion in diagnosis and understanding.

Chapter 3

The Genome's Morbid Anatomy

Throughout the postwar period, medical geneticists sought to increase the number of disorders that could be diagnosed presymptomatically. In the 1960s human cytogeneticists developed the first broad-based approach for advanced diagnosis that relied on chromosomal rather than clinical markers. Over the next decade, human gene mappers promised to expand the set of presymptomatically diagnosable disorders by identifying their chromosomal location. While human cytogeneticists had success in making large chromosomal abnormalities visible under the microscope, in the mid-1970s they believed that many other disease-causing mutations could be visualized only with higher-resolution approaches. Bridging the gap between the few known chromosomal disorders and the many that medical geneticists believed were caused by smaller mutations was a major focus of human cytogenetics in the 1970s and 1980s.

The pursuit of improved delineation and presymptomatic diagnosis of genetic disease, which includes but is not limited to prenatal diagnosis, involved the collaborative work of practitioners in both the laboratory and the clinic.[1] While historically physicians had conducted a morbid anatomy, locating the markers of disease among the bodily organs and tissues of deceased patients, during the postwar period medical geneticists began to look to the human genome, at the microscopically visible level of chromosomes, for new signs and causes of disease. Victor McKusick referred to this practice as examining the genome's morbid anatomy, thus situating it as a medical procedure conducted in a cytogeneticist's laboratory. Like morbid anatomists in previous centuries, medical geneticists sought one-to-one links between visible lesions (now chromosomal) and clinical outcomes. Identifying such correlations required the establishment of distinct categories on each side: dysmorphologists and other physicians identified a pattern of malformation in the clinic that they believed to have a single genetic cause, and human cytogeneticists sought to associate it with a specific chromosomal mutation in the genome's morbid anatomy. This chapter traces the contributions of various medical specialties to the construction of a new clinical disorder called Prader-Willi syndrome and its eventual association with a discrete cytogenetic marker.

Medical geneticists pointed to Prader-Willi syndrome in the 1980s as a representative example of how genetic disorders could be located among the human genome's morbid anatomy using "high-resolution" chromosomal analysis. Prader-Willi syndrome's association with a discrete chromosomal abnormality, medical geneticists noted, facilitated its isolation from other disorders that looked similar in the clinic and thus allowed it to be diagnosed earlier in its life history, providing the opportunity to prevent certain symptoms. The high-resolution cytogenetic analysis used to identify the Prader-Willi mutation could not, however, be made amenable to prenatal diagnosis; thus the disorder remained outside the reach of medical geneticists' prenatal preventive focus in the 1980s. At the same time, the predicted one-to-one relationship between Prader-Willi syndrome and its associated mutation—like many other disorders examined here—proved to be more complex than medical geneticists had anticipated. In the long term, though, the localization of Prader-Willi syndrome to a specific site on chromosome 15 and its inclusion in the growing medical genetics infrastructure for diagnosis did eventually facilitate its prenatal prevention.

Delineation across Specialties

During the first two decades of his life, Albert visited the Zurich Children's Hospital on multiple occasions. When physicians there saw him as an infant, they described Albert as developmentally delayed. His physicians also noted a lack of muscular strength, which they diagnosed as hypotonia, a condition seen in infants who were characterized as "floppy" or "limp." As Albert grew older, he gained muscle strength, but his physicians soon described other concerns, including obesity, an unusually small penis and testicles (hypogenitalism), and mental impairments. While the muscular weakness that Albert had experienced as an infant seemed to have little in common with the difficulties he faced later in childhood, Albert's doctors at the Zurich Children's Hospital came to suspect that they were related. Based on their experience with other patients who, like Albert, were "floppy" as infants but became obese as children, these physicians believed they were seeing a reoccurring developmental pattern, a syndrome, that had never been previously described.[2]

Heinrich Willi, a pediatrician who focused on the neonatal period, had first seen Albert when he was an infant. Albert was being kept for an extended stay in Zurich Children's Hospital because of his muscular weakness, a condition in which Willi specialized. During the 1930s Willi had trained in pediatrics at the University of Zurich under Guido Fanconi, a world-renowned researcher best known for identifying a cancer-causing condition in children now known as Fanconi anemia. Fanconi served as director of the Zurich Children's Hospital from 1929 into the early 1960s and was an important international leader for the field of pediatrics. When Willi first saw Albert, Willi was an assistant medical director in charge of the newborn nursery at Zurich Children's Hospital and was finishing up his dissertation on childhood leukemia. He later became the director of neonatology at the Zurich University Hospital.[3]

Around the age of 12, Albert came to the attention of Andrea Prader and his supervisor, the pediatrician Hans Zellweger, at Zurich Children's Hospital. Prader, a young physician, had also worked under Fanconi, but he had been primarily trained in pediatrics by Zellweger since joining the Children's Hospital in 1947. Prader had also received training in the early 1950s in the United States, where he worked for a short period with L. Emmet Holt Jr., a prominent pediatrician at Bellevue Hospital in New York City, and Lawson Wilkins, a pioneering pediatric endocrinologist at Johns Hopkins, who at this time was supervising the pediatrics residency of David Smith, the founder of dysmorphology. Upon returning to Zurich, Prader again took a job at the Children's Hospital; in 1962 he replaced Fanconi as chair of pediatrics.[4]

Prader and Willi had seen Albert at different stages in his development, when he was suffering from distinct problems. However, as they discussed Albert's condition and that of other patients like him, they identified additional individuals who had experienced a similar pattern of malformation that was characterized by muscular weakness as an infant and by obesity, small genitals, and mental retardation as a child. Their colleague Alexis Labhart, who was an internist at the University of Zurich, also saw Albert, this time as an adolescent. Labhart developed his own curiosity about Albert's condition, because of his interest in endocrinology and metabolics. Together, these three physicians who specialized in different life stages and aspects of human development presented at the Eighth International Conference on Pediatrics in 1956 on Albert and a group of eight other individuals whom they had identified as having a similar condition.[5]

Over its life history, Albert's condition, which became known as Prader-Willi syndrome, crosscut several medical specialties because of its progression from muscular weakness in infancy, to childhood obesity and mental retardation, to adolescent diabetes. The disorder drew the attention of a wide variety of medical specialists, including neonatologists, pediatric and adult endocrinologists, neurologists, and psychiatrists. A common interest in characterizing and identifying the cause of this multifaceted disorder brought these physicians together around the same bedside. Independently they recognized discrete pathological features; together they described a seemingly cohesive developmental pattern of bodily malformation.

A major aim of researchers interested in Prader-Willi syndrome during the following decades was to prevent its most notable features before they started. To do so, medical geneticists believed that they would need to diagnose the syndrome during infancy, before the onset of overeating around age two, which quickly led to childhood obesity and to diabetes later in life. This meant delineating the disorder in the absence of its most prominent clinical features, and instead based on a cytogenetic or biochemical marker.[6] This chapter examines how a chromosomal abnormality associated with Prader-Willi syndrome came to be an exemplar of the one mutation–one disorder ideal of medical genetics and discusses the promising potential application of this model to presymptomatic diagnosis and prevention.

Explaining Childhood Obesity

Postwar physicians often looked to the past for historical accounts of the disease entities they were seeing. They applied a clinical gaze forged in their contemporary medical training to drawings and descriptions from previous eras. These physicians were regularly called upon to make sense of developmental patterns in their patients, and not surprisingly, they tended to extend this mode of analysis into their engagement with art and literature. Undoubtedly, there was a degree of satisfaction to be gained from detecting in a historical account or a painting a disorder that had never previously been recognized. Such retroactive diagnoses also helped to solidify disease categories as real and timeless, rather than contemporary social and medical constructs. They offered physicians a sense that these diseases had always been part of the human experience.

During the early twenty-first century, Prader-Willi syndrome researchers pointed to multiple artifacts as early accounts of the disorder. In the first pages of the 2006 medical text *Management of Prader-Willi Syndrome,* physician Merlin Butler and his colleagues prominently featured two paintings of a young girl who they believed had Prader-Willi syndrome.[7] In 1680 Juan Carreno de Miranda, an artist in the Court of Spanish monarch Charles II, painted the six-year-old girl, known to the court as "La Monstrua." His paintings, one of the girl clothed and the other showing her naked, were on display in 2006 in Madrid's Prado Museum of Art. The authors suggested that these paintings might be the earliest depiction of Prader-Willi syndrome. Similarly, in 1997 physician O. Conor Ward pointed to the mid-nineteenth-century writing of John Langdon Down and made the case that Down had published, in 1864, the first account of a patient with Prader-Willi syndrome. At the time, Down was reporting on a young woman whose obesity he had attempted to manage using strict dietary restrictions, in line with the twentieth-century practice for Prader-Willi syndrome. Down described the patient, by then age 25, as hypotonic, prepubescent, exhibiting distinctive facial features, and having severe learning disabilities.[8] In both of these instances, postwar physicians had looked to historical artifacts and seen a developmental pattern that appeared familiar to them. But such retroactive accounts overlooked the complex history of physicians' examining and making sense of disorders involving childhood obesity.

During the early twentieth century, physicians described multiple developmental patterns that included childhood obesity and small genitals. In 1901 Austrian neurologist Alfred Fröhlich described a disorder in a boy who had regular severe headaches and vomiting, rapid weight gain after age 10, loss of vision in one eye, and unusually small genitals. The boy "did not develop sexually" as a teenager, and Fröhlich noted that "his general appearance remained infantile feminine." He was, however, mentally normal. Fröhlich believed that the boy's condition resulted from the growth of a tumor on his pituitary gland, disrupting his eyesight on one side and leading to abnormal endocrine function. Ultimately, an operation was performed to remove the tumor, leading to general improvement of the boy's headaches and eyesight, as well as the onset of sexual development.[9] In a

1913 follow-up report, Hungarian physician Arthur Biedl, an early endocrinologist, commented that despite other improvements, this patient, by then in his mid-20s, "showed no significant decrease in deposits of fat."[10]

Many physicians in the following decades adopted the eponym "Fröhlich syndrome" as a general term for childhood obesity. The practice of naming disorders after those who were believed to have first described them began in the nineteenth century and continues to the present day. The historical and clinical accuracy of such eponyms have regularly been called into question.[11] In 1939 New York physician Hilde Bruch suggested that the term *Fröhlich syndrome* was frequently misused, partly because Fröhlich's initial report, in an unpublished lecture, was relatively obscure. Bruch reconstructed, in English, Fröhlich's initial report and provided various follow-up accounts of Fröhlich's original patient in the *American Journal of Diseases in Childhood.* Subsequent research on other individuals with similar symptoms had shown that a pituitary gland disruption was probably not the cause of obesity in these cases. Therefore, Bruch argued more broadly that the term *Fröhlich syndrome* was being used much too loosely in diagnosing childhood obesity. Many of the individuals identified as having Fröhlich syndrome showed normal sexual development and often went through puberty early.[12] Bruch also expressed concern about the use of pituitary extracts to treat children haphazardly diagnosed with Fröhlich syndrome, especially because these hormonal remedies were becoming more potent.

In 1922 Beidl, by then the chair of experimental pathology at the University of Vienna, described a developmental pattern in two sisters featuring childhood obesity and hypogenitalism, as in Fröhlich syndrome. However, Beidl also noted that the disorder included mental retardation, the degenerative eye disease retinitis pigmentosa, and polydactyly (extra fingers). Beidl's presentation, at a conference in Vienna, came two years after Georges Bardet, a physician in training, had described the same condition in his University of Paris graduate thesis. Similar developmental patterns, involving childhood obesity and retinitis pigmentosa, could be traced back as far as an 1866 report by English ophthalmologists John Laurence and Robert Moon, who reported on four children in one family who shared these features. In 1925 physicians Solomon Solis-Cohen and Edward Weiss suggested the term *Laurence-Beidl syndrome* for this disorder, acknowledging the part that both ophthalmologists and endocrinologists had played in its delineation. The eponym was widely adopted by physicians, and with it childhood obesity came to be a condition that drew the attention of practitioners from various specialties.[13]

Like many who reported on childhood obesity during the early twentieth century, Solis-Cohen and Weiss presented Laurence-Beidl syndrome as something akin to the developmental pattern that Fröhlich had described, with a few additional features. Similarly, in a 1935 paper on Laurence-Beidl syndrome, one patient was described as having "the complete Fröhlich syndrome, mental retardation, and retinitis pigmentosa." As Bruch later complained, many physicians had taken to using the term *Fröhlich syndrome* as a synonym for childhood obesity, often overlooking the unique characteristics originally described by Fröhlich.[14]

The causes underlying various instances of Fröhlich syndrome remained unclear to physicians during the 1930s and 1940s. A commonplace link between obesity and small genitals suggested a hormonal abnormality, along the lines of what Fröhlich had described. The incidence of mental retardation, retinitis pigmentosa, and other features alongside these issues, however, led some physicians to propose a more complex, potentially hereditary syndrome. This position was bolstered in the case of Laurence-Beidl syndrome by the regular identification of multiple affected individuals in one family.[15] In this era before wide-ranging accounts of recognizable patterns of malformation such as Robert Gorlin and David Smith's texts were available, the identification of multiple abnormal features in a group of similarly affected patients often involved collaboration across medical fields. Sometimes, as in the case of Laurence-Beidl syndrome, this happened through the exchange of published accounts and over long periods of time. At the Zurich Children's Hospital in the 1950s, a new syndrome was delineated through more direct collaboration between physicians from neonatology, pediatric endocrinology, and internal medicine who had all seen the same set of patients over the course of their young lives.

Delineating Prader-Willi Syndrome

In their 1956 report, Prader, Labhart, and Willi distinguished the disorder that they had identified in nine individuals from other similar conditions, by offering a detailed report of its multifaceted life history. At first, they noted, the syndrome resembled other forms of inborn muscular weakness, with significant developmental delay during the first years of life. Later in childhood, the disorder instead appeared to be more like those described by Fröhlich, Laurence, and Beidl. However, an X-ray showed no abnormal pituitary growth, as had been seen in Fröhlich's patient, and there was no sign of retinitis pigmentosa, as would be expected in Laurence-Biedl syndrome. What set the patients of Prader, Willi, and Labhart apart, and suggested a new syndrome, was the sudden transition from muscular weakness to obesity in the second year of life. Based on this reoccurring transformation, the authors argued, "Despite all the similarities, it can be clearly distinguished from the other syndromes mentioned."[16]

Prader, Willi, and Labhart's initial account of this disorder was short, just a few paragraphs, and published in the German-language journal *Swiss Medical Weekly*. The article reflected the specific interests of each author, with its focus on neonatal hypotonia (Willi), pediatric endocrinology (Prader), and metabolic analysis (Labhart). Prader later acknowledged that "at first our paper did not stimulate interest in the medical profession." It was not cited at all during the remainder of the 1950s. The authors, along with Guido Fanconi, also presented their findings at the Eighth International Congress of Pediatrics in Copenhagen in 1956.[17] Even with Fanconi's prominent name attached, the report was not cited in the literature on childhood obesity and hypogenitalism published during the following years. Given the wide diversity of disorders involving these features that had already been described, it seems likely that many physicians were not convinced that the Zurich team had delineated a distinct syndrome.

Having identified five additional patients during the intervening years, Prader and Willi once again presented their cases as a distinct syndrome at the Second International Congress on Mental Retardation in Vienna in 1961. By this time, Prader and Willi had extended their construction of the disorder's life history in both directions, by describing the incidence of minimal fetal movement during pregnancy and the common occurrence of adult-onset diabetes in the teenage years. They also suggested, based on their own experience with 14 individuals and colleagues' unpublished reports of additional cases, that the disorder was likely not all that rare. No cause for the syndrome was yet apparent, though. Prader and Willi noted that cytogenetic analysis in eight patients had not identified any major chromosomal abnormalities, and there was no indication that specific events or exposures during pregnancy were to blame. Recessive inheritance remained a possibility, though no reoccurrence in families had been seen.[18]

Over the ensuing years, Prader and Willi's investigations had an increasing impact on the medical community, as other investigators began to publish their own reports of similar patients. Among the investigators were Bernard Laurance in Derby, England, who described six cases in his article "Hypotonia, Obesity, Hypogonadism and Mental Retardation in Childhood," and Pierre Royer in Paris, who reported in 1963 on the incidence of diabetes among four patients with "le syndrome de Willi-Prader."[19] Additional case reports also emerged in Canada and Sweden. An English doctor, P. R. Evans, in his account of two patients, described the disorder as "resembling Fröhlich's syndrome," with the exception of "presenting at birth with weakness, unwillingness to suck, and under nutrition."[20] Throughout the mid-twentieth century, physicians continued to make reference to disorders involving childhood obesity and small genitals as being like Fröhlich syndrome but with additional features. In the subsequent decades, however, physicians and medical geneticists began to draw finer distinctions among disorders involving childhood obesity and to move away from the vague category of Fröhlich's syndrome, as they sought out and identified potential etiologies among the human chromosomes.

Constructing a Genetic Syndrome

By the end of 1965, clinicians across Europe and North America had described 70 cases of Prader, Labhart, and Willi's syndrome. Many of these investigators agreed that the complex life history of their patients suggested a unique clinical disorder that likely had a single genetic cause. As Prader-Willi syndrome became established as a distinct clinical entity during the mid-1960s, many clinical reports suggested that the disorder might be more common than had been assumed, an impression that was influenced by the report of multiple cases of the syndrome within single institutions.[21] In fact, it was probably the exposure to multiple patients with the same developmental pattern that initially led these physicians to explore the medical literature for an explanation. Those who saw just one patient might overlook the clinical significance of Prader-Willi syndrome's characteristic stages, or they might assume that the common occurrence of these features was coincidental. The growing influence of medical genetics in this era, and the clinical uptake of perspectives and approaches from dysmorphology, undoubtedly influenced some clinicians toward interpreting the developmental stages as reflecting a single malformation pattern.

During the late 1960s, the primary goal of many physicians when publishing on Prader-Willi syndrome was to increase awareness of the disorder and to improve its delineation from other conditions that looked similar during particular moments of their life history. The other conditions included those that featured congenital hypotonia in infancy, such as myotonic dystrophy and spinal muscular atrophy, as well as those associated with childhood obesity, such as Fröhlich and Laurence-Biedl syndromes. Case reports of Prader-Willi syndrome were published in a variety of medical journals that specialized in developmental delay. Notably, in 1969 early and influential dysmorphologists Michael Cohen and Robert Gorlin published a review of multiple patients with Prader-Willi syndrome in the *American Journal of Disability in Childhood.*[22]

In addition to the increasing attention on the part of early dysmorphologists, awareness of Prader-Willi syndrome in clinical communities grew as cytogenetic analysis was becoming increasingly common in the clinic. As a result, Prader-Willi patients' chromosomes were frequently examined for abnormalities. These efforts produced little success. As described earlier, during the mid-1960s cytogeneticists could not distinguish individual human chromosomes, so the clinical value of their analysis was limited. For instance, in 1965 Jean Roget and his colleagues in Grenoble, France, reported on a Prader-Willi patient who had an extra copy of a D group (13–15) chromosome in about 25% of his cells. These researchers also noted a previous report of a child, later diagnosed with Prader-Willi syndrome, who had an extra G group (21, 22, Y) chromosome.[23] A third child, initially reported on in 1963 and later diagnosed as having Prader-Willi syndrome, was also found to have a chromosomal translocation involving the D group. Such findings, however, were sporadic: 43 other Prader-Willi patients that human cytogeneticists examined between 1956 and 1967 were found to have normal karyotypes.[24] While cytogenetic abnormalities were occasionally found in Prader-Willi patients during the 1960s, a lack of specificity about which chromosome was involved limited the clinical value of these findings.

The introduction of banding techniques during the early 1970s offered new opportunities for identifying specific chromosomal abnormalities common to Prader-Willi patients. With the aid of G-banding, British researchers C. J. Hawkey and Alison Smithies were able to more precisely describe a chromosomal abnormality that they identified among the cells of a Prader-Willi patient. The affected individual was a middle-aged man who, despite showing moderate mental retardation, had been diagnosed as having Fröhlich syndrome as a child. He was later diagnosed with Prader-Willi syndrome at age 38. This made his condition amenable to cytogenetic analysis to identify a discrete causative mutation. In 1976 Hawkey and Smithies reported that the patient possessed a translocation involving chromosome 15. Because of its unique G-banding pattern, chromosome 15 had become a distinguishable member of the D group.

Noting that multiple D group abnormalities had been identified previously in Prader-Willi patients, Hawkey and Smithies closed their paper with the suggestion, "It would be tempting to speculate that the number 15 chromosome is involved in its pathogenesis." However, this intriguing hypothesis was tempered by the relative lack of chromosomal anomalies

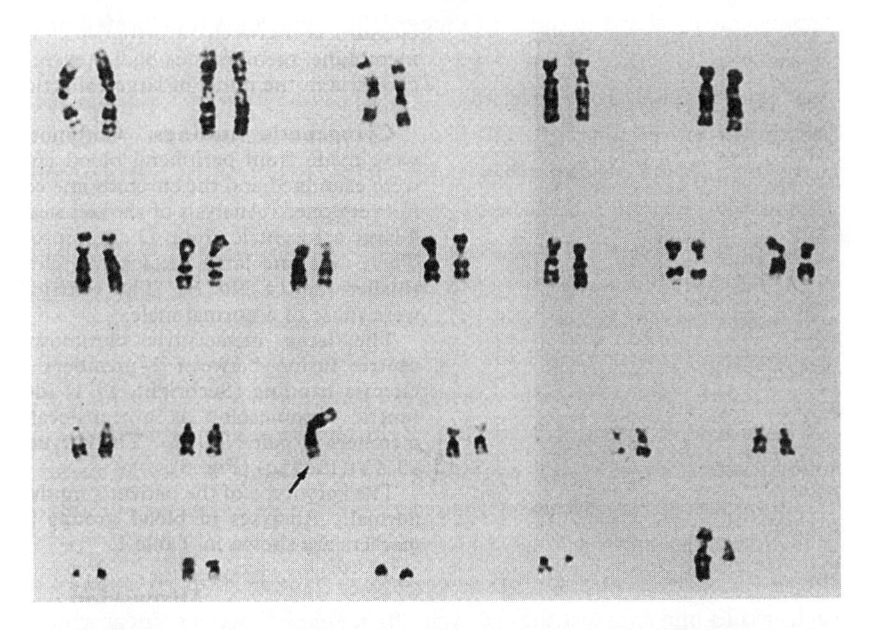

Translocation involving both copies of chromosome 15 identified in a Prader-Willi syndrome patient. *Source:* C. J. Hawkey and A. Smithies, "The Prader-Willi Syndrome with a 15/15 Translocation: Case Report and Review of the Literature," *Journal of Medical Genetics* 13, no. 2 (1976): 154. Reprinted with permission from BMJ Publishing Group Ltd.

that had been reported in Prader-Willi patients overall. Acknowledging the present lack of corroborating evidence, the authors argued, "It would nevertheless be valuable if previously described cases of the Prader-Willi syndrome with abnormal karyotypes were submitted to further analysis with banding."[25] This new cytogenetic technique promised to enhance and clarify the analysis of chromosomal abnormalities among Prader-Willi patients. In addition, during the late 1970s, Hawkey and Smithies's paper drew the attention of human cytogeneticists who were working to develop and apply new laboratory approaches, which later proved integral to locating many disorders within the chromosomal infrastructure of medical genetics, including Prader-Willi syndrome.

High-Resolution Cytogenetics

Much as Herbert Lubs described the fragile X site in families affected by X-linked mental retardation, human cytogeneticists interpreted the chromosomal translocation identified by Hawley and Smithies as demonstrating that the cause of Prader-Willi syndrome was located at a particular place in the human genome, on chromosome 15. In both instances, cytogeneticists found their analysis to be limited by the resolution of chromosomal analysis. Lubs, Hawkey, and Smithies believed that there was a divide between what they could see under the microscope and the causative mutation that was at work in affected patients. The adoption of banding techniques had enhanced the descriptive capabilities of human cytogeneticists, but cytogeneticists continued to express a need to gaze still deeper into the human genome to detect most disease etiologies.

Human cytogeneticists had long understood that inside a living cell, the structure of chromosomes was in a constant state of flux. During most of a human cell's reproductive cycle, the chromosomes were not present in a condensed state and thus could not be made visible under the microscope. As a result, if a cytogeneticist were to examine hundreds of human cells, he would expect to find just a few with visible chromosomes. To increase the yield of visible chromosomes, cytogeneticists used chemicals such as colcemid to arrest the cell cycle during metaphase, a point at which chromosomes were in a condensed state. Using G and R banding techniques, human cytogeneticists were then able make between 300 and 500 bands visible among the 24 human chromosomes. This technique allowed each chromosome to be distinguished, and it aided in the identification of extra, missing, or translocated genetic segments in patients with birth defects and certain genetic disorders.

Beginning in the mid-1970s, University of Minnesota cytogeneticist Jorge Yunis sought to increase the number of bands that could be made visible across the human chromosomes by capturing them in a less condensed state. Yunis believed that such "high-resolution" cytogenetic analysis would facilitate the identification of more subtle chromosomal abnormalities in patients that appeared, using existing approaches, to have normal karyotypes. From here, the novel association of a clinical disorder with the disruption of a particular chromosomal band might point researchers to the location of a causative gene. Importantly, a larger number of visible bands meant that each contained fewer candidate genes. High-resolution chromosomal analysis, Yunis argued, could "begin to bridge the gap between genes and chromosomes in man."[26]

To see more bands, Yunis sought to capture cells at an earlier stage in the cell cycle (prophase or prometaphase), when their chromosomes were less fully condensed. In any culture, a few cells would be found at this stage, but not enough to do reliable cytogenetic analysis (100 cells were considered sufficient). Yunis therefore needed to increase the yield of prophase cells. To do this, he sought to synchronize the cell cycle in a culture by using the chemical amethopterin to block DNA replication, which precedes reproduction. Adding this chemical to cell culture and waiting for 17 hours had the effect of trapping many human cells in the same reproductive stage. When this block was released, all of these cells entered DNA replication and chromosome formation simultaneously. Yunis then found that an additional five hours and 10 minutes was just enough time for many of the synchronized cells to reach prophase. At this point, Yunis would add colcemid to once again disrupt the cell cycle, and he would immediately fix (kill and immobilize) the human cells in preparation for banding and cytogenetic examination.[27]

Yunis's method was more time-consuming and laborious, but it did generate a much higher number of cells in prophase, allowing for high-resolution cytogenetic analysis. Rather than 400 bands, Yunis reported seeing upwards of 1,000 among the 24 human chromosomes. This meant that more abnormalities might be identified in patients whose chromosomes appeared to be normal at a lower resolution. In 1977 Yunis and his colleagues reported that high-resolution cytogenetics had been used to

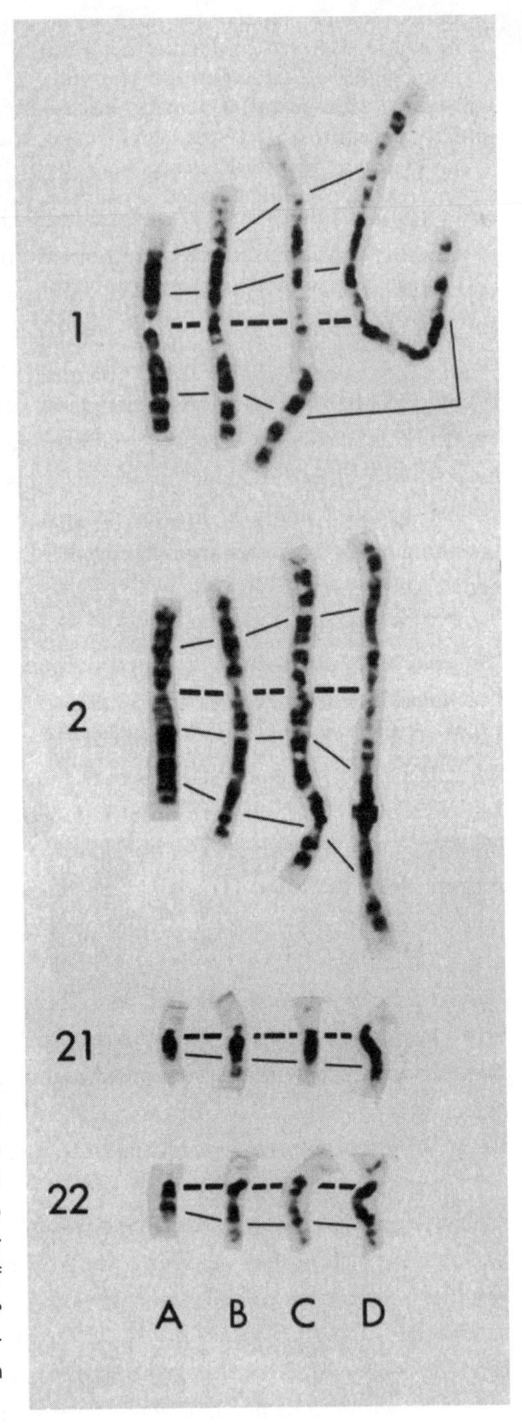

Banded ideograms of high-resolution chromosomes. The accompanying chromosomes demonstrate the increased number of bands compared to previous techniques. *Source:* J. J. Yunis, "High Resolution of Human Chromosomes," *Science* 191, no. 4233 (1976): 1269. Reprinted with permission from AAAS.

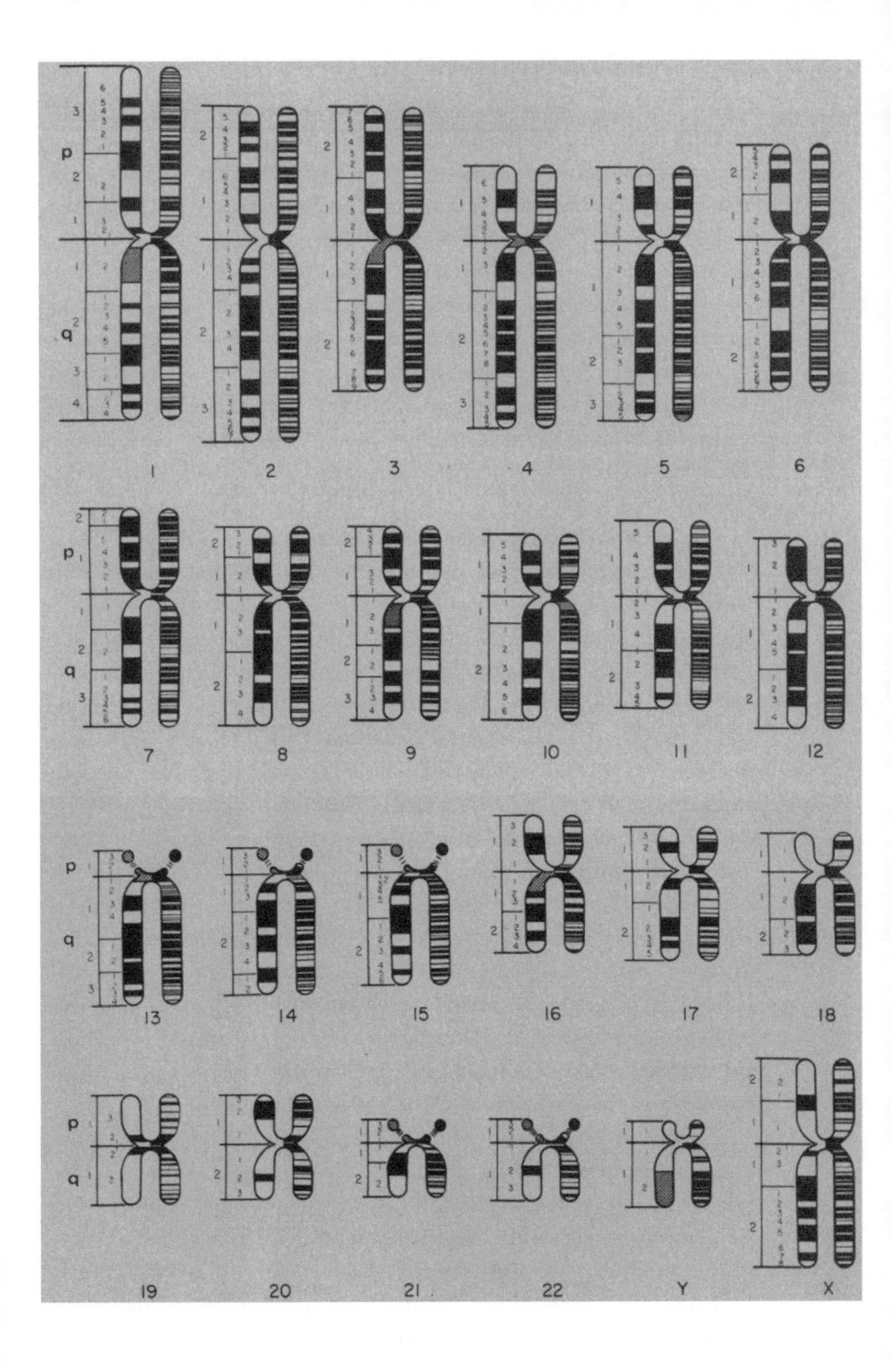

identify minute chromosomal deletions in patients with multiple birth defects as well as in retinoblastoma, a cancer affecting the eyes. High-resolution techniques were also adopted by Vincent Riccari's cytogenetics laboratory at the Baylor College of Medicine during the late 1970s to examine patients with the aniridia-Wilms tumor, which also affected the eye. Application of this technique revealed a consistent deletion in a specific region on the short arm of chromosome 11 in three patients who had previously been found to have normal karyotypes.[28] While Riccardi was initially incredulous, a technician in his laboratory named David Ledbetter soon convinced him that high-resolution cytogenetic techniques could also be used to look for the causative mutation in Prader-Willi syndrome.

Prader-Willi Syndrome in High Resolution

In 1976 Ledbetter was a PhD student in behavioral genetics at the University of Texas, when he came across Hawkey and Smithies's paper on Prader-Willi syndrome in the *Journal of Medical Genetics.* After his first year of course work in Austin, Ledbetter had developed an interest in medical genetics and was spending a few weeks in Galveston to get some laboratory experience at the University of Texas Medical Branch. While there, Ledbetter passed his evenings in the library familiarizing himself with the recent literature in human cytogenetics. He was fascinated by Hawkey and Smithies's recent identification of a translocation in a Prader-Willi patient and by the idea that an abnormality in chromosome 15 might cause the disorder. As Ledbetter continued his PhD program in Austin, his adviser encouraged him to continue visiting medical genetics laboratories, including the clinic of Jose Louro in San Antonio. During one of his visits, a young girl was seen in Louro's clinic who showed the classical features of Prader-Willi syndrome and also possessed a chromosome 15 translocation. Ledbetter asked Louro if he was familiar with Hawkey and Smithies's paper. He was not, so Ledbetter forwarded it to him.[29]

Having developed a keen interest and some basic skills in human cytogenetics during his early years of graduate school, Ledbetter was invited to spend a semester during 1978 working in the laboratory of the well-known cytogeneticist T. C. Hsu at the M. D. Anderson Medical Center in Houston. What was supposed to be a few months turned into a three-year stay in Hsu's laboratory, during which Ledbetter completed his dissertation on evolutionary cytogenetics in primates. Since he had no funding to support him during his time in Houston, he also took a position in Riccardi's cytogenetics laboratory, which proved to be a fruitful setting for Ledbetter to pursue his interest in the genetic basis of Prader-Willi syndrome.[30]

Riccardi's report on the chromosomal basis of the aniridia-Wilms tumor was published in 1978, the same year Ledbetter joined his laboratory as a part-time technician. By this time, two additional research groups had pointed to a chromosome 15 translocation in Prader-Willi syndrome, adding further support to the hypothesis of Hawkey and Smithies. As Italian medical geneticist Marco Fraccaro stated in his 1977 article, "The finding of three or possibly four cases of 15/15 translocation associated with Prader-Willi syndrome leaves little doubt that chromosome 15 is involved in the pathogenesis of this syndrome."[31] Over the next few years, additional reports linking Prader-Willi syndrome and a chromosome 15 translocation were published, leading to the suggestion in 1979 that about 10% of patients with the disorder showed a D group chromosomal abnormality.[32] As he worked in Riccardi's laboratory, Ledbetter learned high-resolution chromosome analysis and contributed to its application in the study of retinoblastoma.[33] Among the tissue samples that the laboratory received from patients diagnosed with various other disorders, however, very few were found to possess the minute deletions of genetic material that high-resolution cytogenetics had been developed to identify.[34]

In 1979 a physician-referred tissue sample arrived that was taken from a young girl who had been clinically diagnosed with Prader-Willi syndrome. According to Ledbetter, Riccardi had little expectation that the sample would show any cytogenetic abnormality, despite mounting reports of chromosome 15 translocations in similarly affected individuals. At this time, clinical cytogeneticists, already burdened by a large caseload, were often frustrated by physicians who unknowingly sent samples from patients with disorders that most medical geneticists regarded as being caused by single-gene mutations, rather than chromosomal abnormalities. Initially it seemed as if Riccardi's inclination in this case was correct: the laboratory found a normal karyotype. But when Ledbetter was able to take a closer look at this child's chromosomes, he "was struck by an apparent length discrepancy between the two chromosome 15 homologs, and the fact that the shorter homolog appeared 'paler' just below the centromere." This finding suggested to him that there might be a small deletion present on chromosome 15.[35]

To further investigate Ledbetter's hypothesis, Riccardi arranged for another sample from this Prader-Willi patient to be prepared for high-resolution chromosomal analysis. Ledbetter was able to make the deletion appear even more distinctly with high-resolution analysis, which confirmed for him that this patient possessed a small deletion on chromosome 15. One finding alone did not definitively demonstrate that a chromosome 15 deletion caused Prader-Willi syndrome. For Ledbetter and Riccardi, the next step was to acquire cells from 5 or 10 more patients clinically diagnosed with the disorder, along with some normal control samples, to determine whether many Prader-Willi patients possessed the same deletion. Blood samples were acquired from John D. Crawford, a pediatric endocrinologist at Massachusetts General Hospital, for a blinded study. Ledbetter did not know which sample was which, but once again he was able to distinguish those patients with Prader-Willi syndrome from the other control samples using high-resolution cytogenetic analysis. The identification of the same chromosomal deletion in four out of five patients that Crawford had clinically diagnosed with Prader-Willi syndrome demonstrated more authoritatively that the disorder was caused by a common mutation.[36]

The next year, Ledbetter was invited to present his findings at the plenary session of the American Society of Human Genetics meeting. He and Riccardi published a report of their findings in the *New England Journal of Medicine* soon thereafter.[37] The chromosomal images displayed in their paper made clear to the casual observer how subtle the distinction was between a normal chromosome 15 and one with a deletion at the banding nomenclature location 15q11–13. In these photographs, nature did not speak for itself. Indeed, as with the delineation of malformation patterns in the clinic by dysmorphologists, the detection of chromosomal abnormalities in Prader-Willi syndrome patients depended on the standardized techniques of human cytogenetics and the judiciously developed trained judgment of laboratory observers.[38]

Human cytogeneticists had long understood chromosomes, their work objects, to be unruly biological entities. With the adoption of high-resolution approaches, cytogeneticists were increasingly looking for clinically relevant abnormalities that were at the edge of their visual perception.

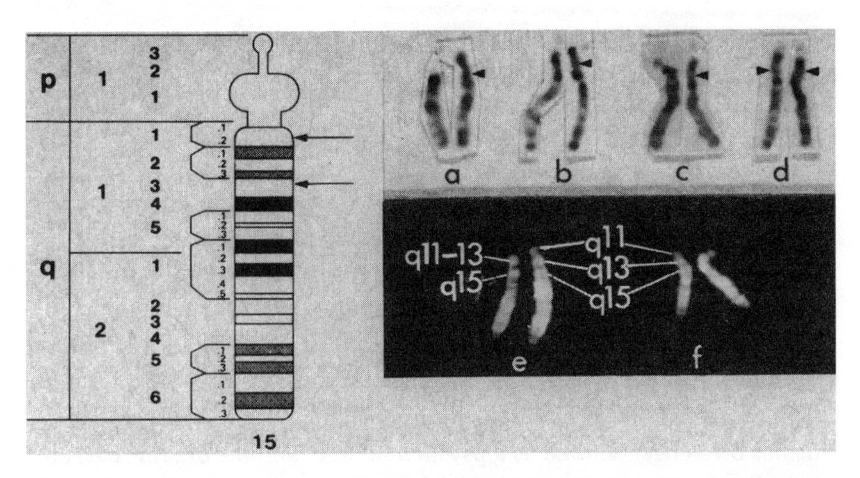

High-resolution ideogram and chromosomes showing the detection of a microscopically visible deletion on chromosome 15 in Prader-Willi syndrome patients. *Source:* D. H. Ledbetter et al., "Deletions of Chromosome 15 as a Cause of the Prader-Willi Syndrome," *New England Journal of Medicine* 304, no. 6 (1981): 327. Copyright © Massachusetts Medical Society. Reprinted with permission from Massachusetts Medical Society.

Blinded studies such as Ledbetter's were frequently employed in these cases to verify that apparent cytogenetic findings were consistent across practitioners and samples. Human cytogeneticists accepted that there was an inevitable, but manageable, degree of subjectivity inherent in their craft. As historians of science Lorraine Daston and Peter Galison compellingly demonstrated, the pursuit of objectivity in science has had much to do with changing norms among researchers about what forms of subjectivity were acceptable, and what forms were seen as "*dangerously* subjective."[39] Cytogeneticists were willing to accept that the visibility of a given mutation could be somewhat variable, requiring a subjective inference that it was present in every cell, even when if it could not always be made visible. They were concerned, however, that trained judgment alone could not overcome the potential for confirmation bias when examining the chromosomes of a patient for whom they already had a clinical diagnosis in mind. By doing a blinded study, human cytogeneticists reassured themselves that they had actually made visible a real but subtle mutation, independent of external factors impacting their subjective expectations about its presence. But, as a later section of this chapter details, the challenge of visualizing a chromosomal deletion in many Prader-Willi patients was a barrier to its broader diagnostic value.

Prader-Willi Syndrome as Exemplar

The association of Prader-Willi syndrome with a visible deletion on chromosome 15 was widely reported in the medical literature during the early 1980s. In a 1981 paper, Victor McKusick referenced Ledbetter's finding, suggesting that having a chromosomal marker for Prader-Willi syndrome facilitated its delineation from the vague category of Fröhlich syndrome as well as other, more stigmatizing, psychiatric explanations. He pointed out, "Like other problems lumped together as eating disorders, this has often been viewed as a psychiatric state and the organic basis revealed by the chromosomal aberration has been in my experience a relief to the families of the afflicted."[40] The identification of a chromosome 15 deletion among Prader-Willi patients facilitated a clearer diagnosis and, even more importantly, a straightforward explanation for families. Like a distinct pattern of malformation in dysmorphology, the chromosomal deletion demonstrated that the disorder had occurred randomly, meaning no one was to blame.

McKusick further highlighted the significance of the Prader-Willi deletion in medical genetics by including the disorder on his regularly updated "morbid anatomy of the human genome" diagrams beginning in 1981.[41] These chromosome-level depictions, which resembled his human gene maps, showed where in the human genome the suspected etiologies of many genetic disorders were located. Like the gene maps, the morbid anatomy figures were built on the standardized set of banded chromosomal ideograms developed at the 1971 Paris conference. McKusick's conception of the genome's morbid anatomy reflected the reductive one mutation–one disorder ideal of postwar medical genetics. His diagrams located the causes of disease at discrete chromosomal locations in much the same way that eighteenth-century morbid anatomists had associated diseases with organ or tissue lesions during postmortem examination.

With high-resolution techniques, human cytogeneticists in the late 1970s began to associate disorders with genetic deletions that were just one or a few chromosomal bands in size. The precision with which mutations for aniridia-Wilms tumor, retinoblastoma, and Prader-Willi syndrome had been located was a significant reflection, for McKusick, of what medical genetics sought to achieve. In comparing McKusick's gene maps and morbid anatomy diagrams, his former student Reed Pyeritz commented, "You have the same ideogram, but now you have phenotypes/diseases, and [McKusick] said, 'this is what a geneticist does, this is the morbid anatomy.'"[42]

In 1982 medical geneticist Charles Scriver similarly pointed to the demonstration of Prader-Willi syndrome's cytogenetic basis as representative of improvements in examining the genetic cartography and "neo-Vesalian anatomy" of the human genome. Scriver noted that by using high-resolution techniques, cytogeneticists had been able to identify the "chromosomal addresses" of Prader-Willi syndrome, retinoblastoma, and the aniridia-Wilm's tumor. Looking forward, Scriver argued that an awareness of the chromosomal addresses of disease would "be used systematically to diagnose the chromosomal phenotype prospectively. This represents an interesting development in genetic counseling."[43] Moving beyond the few disorders, including Down syndrome, that were caused by large chromosomal abnormalities, Scriver suggested that high-resolution cytogenetics might offer new options for presymptomatic and prenatal prevention.

Throughout the 1980s, medical geneticists pointed to high-resolution cytogenetics as playing a significant role in identifying the location of disorders in the genome's morbid anatomy. In the 1985 edition of the textbook *An Introduction to Medical Genetics,* John Fraser Roberts and Marcus Pembrey pointed to the value of high-resolution cytogenetics: "Chromosomal defects, particularly deletions, are being found in an increasing number of conditions. For example, about half the cases of Prader-Willi syndrome . . . reveal a deletion or some other defect of 15q11–12."[44] Prader-Willi syndrome was an exemplar of this success and was frequently pointed to, including in Gerald Stine's 1989 textbook *The New Human Genetics,* which stated, "With the use of high-resolution banding techniques, a growing number of interstitial deletions have been found and associated with diseases of previously unknown cause. . . . Prior to high-resolution banded chromosome identification, the Prader-Willi syndrome was thought to be due to a recessive mutation or multifactorial inheritance."[45] Reflecting McKusick and Scriver's vision of what high-resolution cytogenetics contributed to medical genetics, Stine noted that the technique had helped to transform Prader-Willi syndrome from a set of symptoms of

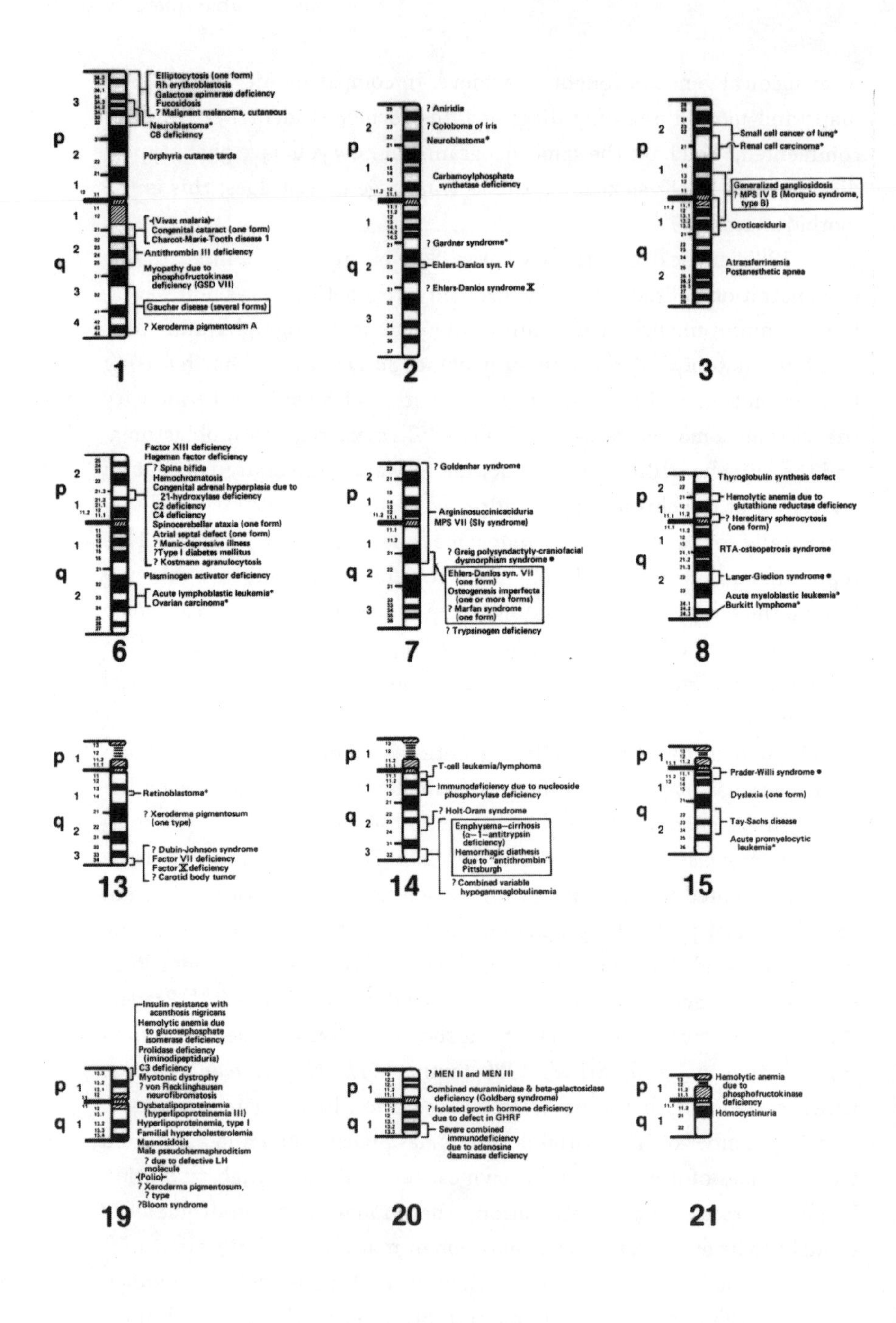
Elliptocytosis (one form)
Rh erythroblastosis
Galactose epimerase deficiency
Fucosidosis
? Malignant melanoma, cutaneous
Neuroblastoma*
C8 deficiency
Porphyria cutanea tarda
(Vivax malaria)
Congenital cataract (one form)
Charcot-Marie-Tooth disease 1
Antithrombin III deficiency
Myopathy due to phosphofructokinase deficiency (GSD VII)
Gaucher disease (several forms)
? Xeroderma pigmentosum A
1
? Aniridia
? Coloboma of iris
Neuroblastoma*
Carbamoylphosphate synthetase deficiency
? Gardner syndrome*
Ehlers-Danlos syn. IV
? Ehlers-Danlos syndrome X
2
Small cell cancer of lung*
Renal cell carcinoma*
Generalized gangliosidosis
? MPS IV B (Morquio syndrome, type B)
Oroticaciduria
Atransferrinemia
Postanesthetic apnea
3
Factor XIII deficiency
Hageman factor deficiency
? Spina bifida
Hemochromatosis
Congenital adrenal hyperplasia due to 21-hydroxylase deficiency
C2 deficiency
C4 deficiency
Spinocerebellar ataxia (one form)
Atrial septal defect (one form)
? Manic-depressive illness
?Type I diabetes mellitus
? Kostmann agranulocytosis
Plasminogen activator deficiency
Acute lymphoblastic leukemia*
Ovarian carcinoma*
6
? Goldenhar syndrome
Argininosuccinicaciduria
MPS VII (Sly syndrome)
? Greig polysyndactyly-craniofacial dysmorphism syndrome ●
Ehlers-Danlos syn. VII (one form)
Osteogenesis imperfecta (one or more forms)
? Marfan syndrome (one form)
? Trypsinogen deficiency
7
Thyroglobulin synthesis defect
Hemolytic anemia due to glutathione reductase deficiency
? Hereditary spherocytosis (one form)
RTA-osteopetrosis syndrome
Langer-Giedion syndrome ●
Acute myeloblastic leukemia*
Burkitt lymphoma*
8
Retinoblastoma*
? Xeroderma pigmentosum (one type)
? Dubin-Johnson syndrome
Factor VII deficiency
Factor X deficiency
? Carotid body tumor
13
T-cell leukemia/lymphoma
Immunodeficiency due to nucleoside phosphorylase deficiency
? Holt-Oram syndrome
Emphysema—cirrhosis (α—1—antitrypsin deficiency)
Hemorrhagic diathesis due to "antithrombin" Pittsburgh
? Combined variable hypogammaglobulinemia
14
Prader-Willi syndrome ●
Dyslexia (one form)
Tay-Sachs disease
Acute promyelocytic leukemia*
15
Insulin resistance with acanthosis nigricans
Hemolytic anemia due to glucosephosphate isomerase deficiency
Prolidase deficiency (iminodipeptiduria)
C3 deficiency
Myotonic dystrophy
? von Recklinghausen neurofibromatosis
Dysbetalipoproteinemia (hyperlipoproteinemia III)
Hyperlipoproteinemia, type I
Familial hypercholesterolemia
Mannosidosis
Male pseudohermaphroditism ? due to defective LH molecule
(Polio)
? Xeroderma pigmentosum, ? type
?Bloom syndrome
19
? MEN II and MEN III
Combined neuraminidase & beta-galactosidase deficiency (Goldberg syndrome)
? Isolated growth hormone deficiency due to defect in GHRF
Severe combined immunodeficiency due to adenosine deaminase deficiency
20
Hemolytic anemia due to phosphofructokinase deficiency
Homocystinuria
21

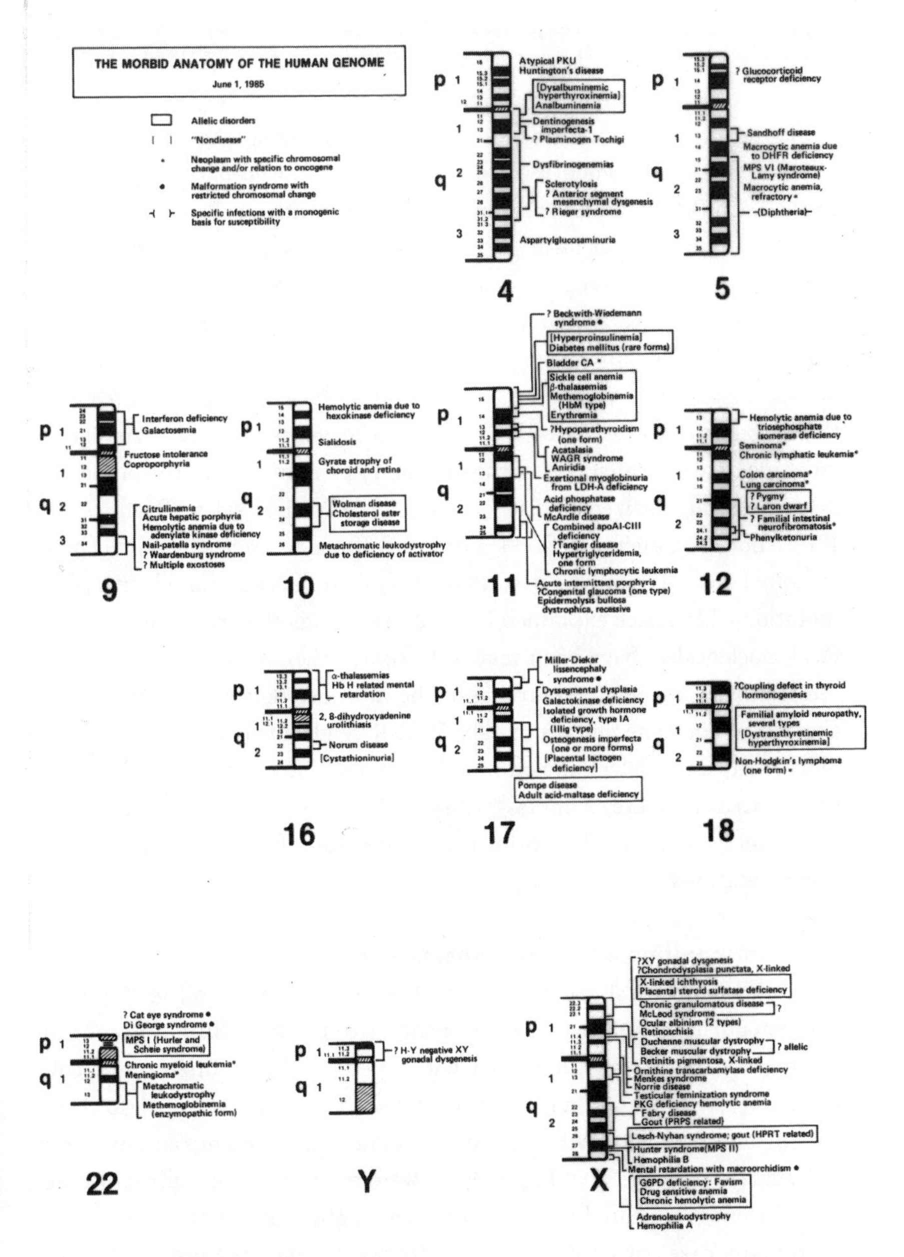

Updated version of the morbid anatomy of the human genome diagram produced by Victor McKusick in 1985. *Source:* Reprinted with permission from the Alan Mason Chesney Medical Archives of the Johns Hopkins Medical Institutions.

unknown origin into a discrete genetic disorder, and it promised to do the same for other combinations of symptoms.

The genomic pinpointing of a disorder following the identification of a common "microdeletion" in clinically diagnosed patients, such as 15q11–13 in Prader-Willi syndrome, fit perfectly with the broadly held one mutation–one disorder ideal. Medical geneticists were enthusiastic about identifying complex syndromes that could be linked to specific chromosomal abnormalities, which might in turn eventually lead them to a causative gene. Historically, they had assumed that single-gene disorders would not be diagnosable cytogenetically, because the mutations that caused them would be too small to see. Lionel Penrose pointed to this when distinguishing between "mistakes of the imaginary printer" and of "the binder," a distinction popularized by historian of science Daniel Kevles.[46] In the early 1980s, however, microdeletion syndromes offered examples in which chromosomal analysis could be used to identify disorders associated with gene mutations. McKusick explained in 1982, "Deletions at a grosser level [than DNA nucleotides] have been related to some other abnormalities and we are seeing an interesting blurring of the interface between chromosomal aberration and gene mutations. The WAGR syndrome [aniridia-Wilms tumor] on 11p, retinoblastoma on 13q, the Prader-Willi syndrome in the paracentric area of 15q are examples."[47] Just as Yunis had hoped, high-resolution cytogenetics seemed to be closing the gap between chromosome and nucleotide mutations.[48]

Interpreting a Chromosomal Deletion

While medical geneticists attached great promise to the 15q11–13 deletion during the 1980s, its relationship with Prader-Willi syndrome did not prove to be as clearcut as they had hoped. In 1982 Ledbetter and colleagues followed up on their initial report with a broader study of 40 clinically diagnosed Prader-Willi syndrome patients. Their findings among these individuals were less striking: only 19 were found to have a chromosome 15 deletion. It appeared that a chromosomal abnormality was associated with many cases of Prader-Willi syndrome, but perhaps not all. Microscopically identifying a chromosomal deletion at the banding location 15q11–13, near the centromere on the long arm, seemed to represent a reliable means of diagnosing Prader-Willi syndrome, but the absence of this deletion could not be used to exclude the disorder.[49] Additional research groups reported similarly disconcerting findings in the early 1980s. Merlin Butler, a physician who was completing his PhD in medical genetics at Indiana University, found the same chromosome deletion in 7 of 14 clinically diagnosed Prader-Willi patients.[50] Around the same time, Jean-François and Marie-Genevève Mattei in Marseille, France, were also doing cytogenetic workups on Prader-Willi patients. In 1983 they reported finding the deletion in 12 of 17 cases. Despite the varied expression of chromosome 15 abnormalities, even in their own patients, the Matteis argued in a review published the next year that this category of cytogenetic mutations was common enough in Prader-Willi syndrome to make the disorder "indissociable" from the 15q11–13 region.[51]

Even as some medical geneticists pointed to Prader-Willi syndrome as an exemplary microdeletion syndrome, many researchers who were familiar with the disorder acknowledged that its association with a chromosomal deletion was not so simple. During the 1980s the group of chromosome 15 abnormalities identified among Prader-Willi patients was diverse; it included duplications and inversions of genetic material in the 15q11–13 region in addition to the deletions that Ledbetter had described. These findings led some medical geneticists to ponder whether chromosome 15 abnormalities were the result of Prader-Willi syndrome, rather than its cause.[52] Similar suggestions were also made about the fragile X site in the 1980s, demonstrating the continued uncertainty at this time about the nature of causal mechanisms linking chromosomal abnormalities and clinical outcomes. This was an issue that Yunis, McKusick, and other medical geneticists had hoped that increasingly high-resolution analysis would solve. However, even in the exemplary case of Prader-Willi syndrome, uncertainty remained.

While some medical geneticists expressed great confidence in the link between Prader-Willi syndrome and 15q11–13 deletions, they still needed to account for the significant variability in cytogenetic findings across clinically diagnosed patients. Ledbetter and colleagues suggested possible explanations. One was that deletions existed in all patients but were too small to be seen in half of the cases, even with high-resolution techniques. Other plausible explanations included the possibility that Prader-Willi syndrome had multiple causes or was being diagnosed in individuals who had similar but genetically distinct disorders.[53] The ideals of medical genetics, as outlined by McKusick, favored the latter explanation. As he expressed it in his introduction to *Mendelian Inheritance in Man* (*MIM*), "They are either the *same* disease, if they are based on the same mutation, or they are *different* diseases."[54] For McKusick and many medical geneticists, a genetic means of delineation was always more reliable than a clinical one.

In the visually messy world of cytogenetics, the variable expression of a chromosomal abnormality in clinically similar patients was a regular occurrence. Soon after Jerome Lejeune associated Down syndrome with the presence of an extra chromosome, an individual also diagnosed clinically with the disorder was found to have the normal complement of 46. This finding was eventually explained by the identification of another chromosomal abnormality, which masked the presence of the extra, Down-syndrome-causing, chromosome.[55] Similarly, in fragile X syndrome the variable visibility of the disorder's chromosomal marker and namesake was a source of uncertainty. Although medical geneticists looked to chromosomal abnormalities for new genetic insights and improved diagnosis during the 1970s, their visual interpretation was often a complex undertaking, involving complicated cytogenetic approaches and requiring significant expertise in examining chromosomes under the microscope. As it stood in the mid-1980s, the absence of a 15q11–13 deletion in half of all clinically diagnosed Prader-Willi patients was a barrier to reliable diagnosis. Additional findings in the following years only added to this uncertainty, complicating initial efforts by medical geneticists to label the disorder as an exemplar of the one mutation–one disorder ideal and as a target for presymptomatic prevention.

The Molecular Turn

A 1988 review by Ledbetter and physician Suzanne Cassidy of 195 published Prader-Willi cases that had been examined using high-resolution cytogenetics found that 59.5% showed a chromosome 15 microdeletion, 37.0% had normal chromosomes, and 3.6% possessed some other chromosome 15 abnormality. The exact role of chromosome 15 abnormalities in causing this disorder remained a matter of debate among cytogeneticists, but the pragmatic clinical value of this marker, the authors noted, was not in question. Throughout their report, which was published in an edited volume aimed at clinicians, Ledbetter and Cassidy highlighted the importance of doing high-resolution cytogenetic analysis despite the potential for uncertain findings. While the absence of a chromosome 15 microdeletion did not exclude Prader-Willi syndrome, its presence was sufficient to make a positive diagnosis. Because the disorder's life history played out in stages, it could not be diagnosed clinically with certainty until after infancy, when its second characteristic phase began. High-resolution cytogenetic analysis, however, could provide an immediate diagnosis in many cases. Ledbetter and Cassidy noted that earlier diagnosis had many benefits for parents, who were seeking the peace of mind that a clear answer could provide, and for the affected infant, who might be spared additional testing and could be put on a regimented diet before the onset of overeating and obesity.[56]

The potential for prenatal diagnosis of Prader-Willi syndrome was another point of discussion during the mid-1980s that Ledbetter and Cassidy addressed. Medical geneticists believed that Prader-Willi syndrome was caused by a de novo mutation in the vast majority of cases. Thus, the disorder did not run in families, meaning there was no reason to believe that any one pregnancy was at increased risk for it. In addition, the cell synchronization necessary for high-resolution cytogenetic analysis proved to be much more difficult in amniocytes than in blood cells, limiting the potential for identifying microdeletions prenatally. Practically speaking, therefore, prenatal diagnosis of Prader-Willi syndrome was infeasible at the time. Ledbetter and Cassidy, among others, argued that attempts to diagnose Prader-Willi syndrome prenatally would only put increased strain on cytogenetics laboratories, without offering any clinical value.[57] Preventive options for the disorder in the mid-1980s were primarily oriented toward reducing its impact through early clinical intervention following postnatal diagnosis. High-resolution cytogenetic analysis had helped medical geneticists to pinpoint the location of Prader-Willi syndrome within the genome's morbid anatomy and chromosomal infrastructure for diagnosis but did not contribute to its widespread prenatal detection and prevention at that time.

Even in the postnatal context, cytogeneticists were frequently unable to confidently confirm or rule out a 15q11–13 deletion. A 1986 study of four cytogenetic laboratories using high-resolution analysis to diagnose Prader-Willi syndrome found that false positive and false negative results were rare. However, there were still many instances of uncertainty over whether a 15q11–13 deletion was present in a particular sample.[58] Many

Prader-Willi researchers were saying that increasing the accuracy of cytogenetic diagnosis, beyond the 50% or so of patients who possessed a clearly visible microdeletion, would require the development of a new type of test for the disorder. Assuming that many of the Prader-Willi patients who appeared to have normal chromosomes did in fact possess a microdeletion, one that was too small to be seen, medical geneticists believed that molecular probes specific to the deletion region could offer improved diagnostic certainty by circumventing the limited resolution of cytogenetic analysis.[59]

Prader-Willi syndrome specialists also suggested that molecular diagnosis might eventually aid in the identification of a specific gene or genes that caused the disorder. Because the 15q11–13 microdeletion was often large enough to be made visible under the microscope, medical geneticists believed that this mutation likely resulted in the loss of many genes. They assumed that the function of most of these genes was unaffected by the absence of one copy, since the same gene on the other copy of chromosome 15 would make up for its loss. Instead, researchers believed that just one or a few genes, which were not fully complemented following the chromosome 15 deletion, caused Prader-Willi syndrome.[60]

In a 1986 *Journal of Pediatrics* article, University of Pennsylvania clinical geneticist Roy Schmickel proposed a new category of disorders that he coined "contiguous gene syndromes." This classification resulted directly from the recent identification of multiple disorders associated with visible microdeletions, including retinoblastoma, aniridia-Wilms tumor, Prader-Willi syndrome, and DiGeorge syndrome, the topic of chapter 5 in this book. Contiguous-gene syndromes represented a middle ground between disorders caused by a large chromosomal abnormality, like Down syndrome, and those caused by a single-gene mutation. Schmickel suggested that the complex and variable clinical features of contiguous-gene syndromes resulted from the loss of two or more genes because of a chromosomal microdeletion. Addressing Prader-Willi syndrome specifically, he cited a report from 1983 of a child with "expanded" Prader-Willi syndrome, who showed additional malformations involving the heart, eyes, and palate. Notably, this child was also found to have an unusual chromosome 15 mutation, involving additional deleted bands, which Schmickel believed led to the loss of one or more additional clinically relevant genes.[61]

Schmickel explained that in contiguous-gene syndromes the causative genes were likely to have nothing in common aside from their proximity. "These genes may be quite independent and no more related than apples and Appalachian Mountains; the loss of an encyclopedia page could remove both entries."[62] While the clinical features of these syndromes resulted from the loss of two or more genes, the impact of each was considered independently. From a research perspective, the most valuable aspect of contiguous-gene syndromes was in the localization of individual genes and the determination of their function. "If we know where a gene is," Schmickel said, "we can find out what is defective and ultimately how it should work."[63] Contiguous-gene syndromes represented another example of what McKusick referred to as "photographic negatives from which a positive picture of man's genetic constitution can be made."[64] Analysis of this new category of genetic disorders thus promised to have both clinical and basic research value, producing knowledge about both the normal and the pathological human genome. Such information proved especially productive in the case of Prader-Willi syndrome.[65]

An Alternative Diagnosis?

Although 15q11–13 deletions were not identified in every Prader-Willi patient during the first half of the 1980s, when this mutation was seen microscopically, medical geneticists believed it was a reliable diagnostic indicator of Prader-Willi syndrome. And despite some variability in clinical expression among Prader-Willi patients, it was widely held that the loss of a specific portion of genetic material on chromosome 15 directly caused the disorder. In 1987, however, high-resolution cytogenetic analysis performed by Ellen Magenis and her colleagues at the Oregon Health Sciences Center called these assumptions into question. Magenis had identified two unrelated girls, ages 5 and 15, who she believed were affected by the same condition, characterized by severe mental retardation, seizures, and jerky movements. Using high-resolution techniques, Magenis identified a 15q11–13 chromosomal deletion in cells from each of these girls. Neither showed any signs of obesity, infant muscular weakness, or other clinical features associated with Prader-Willi syndrome.[66]

In June 1987 Magenis brought her findings to the Prader-Willi Syndrome Association (PWSA) annual conference in Houston. PWSA was a national parent support group founded in 1979 to benefit affected individuals and their families. Beginning in 1985, Prader-Willi syndrome researchers had started holding a one-day professional meeting and research update as part of the larger conference. In 1987 the research arm of the conference was held at the Baylor College of Medicine, where David Ledbetter ran a human cytogenetics laboratory. More than 20 papers were presented, with 47 researchers in attendance. The main PWSA conference continued over the next three days, with more than 500 attendees, including 102 individuals with Prader-Willi syndrome.[67] During the research meeting, Magenis presented images of her two patients. A physician named Charlotte Lafer, who was in the audience, identified the girls based on their clinical appearance as having a distinct clinical disorder called Angelman syndrome.[68]

The conference report from the 1987 meeting stated, "Presentation of these unique cases stimulated discussion and some frustration at the emerging complexity of the phenotype-karyotype correlation in PWS and chromosome 15."[69] Many of the participants agreed that only molecular analysis of the 15q11–13 chromosomal region would bring about the resolution of this uncertainty. Such work was already under way in the laboratory of Samuel Latt at the Boston Children's Hospital. Latt's group had recently reported on additional instances of the 15q11–13 deletion in patients who did not show the clinical features of Prader-Willi syndrome. In a paper published earlier in 1987, first author Lawrence Kaplan identified two patients with this genetic mutation who had been diagnosed with Williams and Angelman syndrome and a third with the deletion who had some, but not all, of the classical clinical features of Prader-Willi syndrome.[70]

These new findings from Oregon and Massachusetts called into question the assumption of a one-to-one correlation between Prader-Willi syndrome and a deletion at the chromosomal address 15q11–13. The same year, Ledbetter and pediatrician Frank Greenberg, at the Baylor College of Medicine, also reported on two individuals who possessed the 15q11–13 deletion but who showed developmental patterns that were not suggestive of Prader-Willi syndrome. Despite these complications, Ledbetter and Greenberg continued to treat Prader-Willi syndrome and other disorders associated with this region as if they had discrete and localized genetic etiologies. The authors closed their paper by noting that "the Angelman syndrome or phenotype might represent a separate locus within the same region of 15q11–q13. These hypotheses will require further delineation and clarification on a molecular level."[71] Ledbetter and Greenberg, like many of their colleagues, believed that confusion over the chromosome 15 microdeletion's role in various distinct clinical disorders demonstrated that high-resolution cytogenetic analysis had reached its diagnostic limits. They were certain that Prader-Willi and Angelman syndromes, which appeared to be very distinct clinically, were not the same genetic disorder and therefore were caused by different mutations. Medical geneticists expressed confidence that molecular analysis would uncover a straightforward resolution to this confusing situation and would uphold the one mutation–one disorder ideal. The subsequent adoption of molecular approaches in human cytogenetics during the 1980s and 1990s are explored in the next chapter.

Conclusions

During the late 1970s and early 1980s, medical geneticists presented high-resolution cytogenetic analysis as an innovative approach for locating discrete genetic disorders within the human genome's morbid anatomy. Human cytogeneticists pointed to the technique as a means "to bridge the gap" between chromosomal markers and molecular genes, and Prader-Willi syndrome was identified as a proof-of-principle and exemplar of these promises. For medical geneticists, the Prader-Willi microdeletion demonstrated that, if properly refined, high-resolution cytogenetics could become a technique that linked the examination of Penrose's mutually exclusive mistakes of the "printer" and the "binder," making chromosomal analysis an important first step in locating and molecularly characterizing disease-causing genes.[72]

As it turned out, high-resolution cytogenetic analysis never provided the level of enhanced vision or clarity that its promoters had predicted. In retrospect, Ledbetter concluded, "It was interesting because, for a while, most people talked about high resolution [cytogenetics] as, 'now we have this tool that we could go out and search for new microdeletions,' and in fact none of the so called microdeletion syndromes were discovered using high resolution. They were all discovered by the patient who had a bigger, obvious deletion or unbalanced translocation, and then we had a tool to screen for other patients for smaller deletions."[73] Even with high-resolution banding techniques, microdeletions were almost impossible to find if a cytogeneticist did not already know where among the chromosomes to look. For this reason, along with technical difficulties, cytogeneticists did not regard high-resolution analysis as being amenable to prenatal diagnosis, which was a major focus of medical geneticists in the 1980s.[74] Indeed, a cytogeneticist was unlikely to identify an unexpected microdeletion in a fetal sample, and even if she did, she most likely lacked any basis for interpreting its clinical significance or impact. As later chapters describe, ongoing growth of the chromosomal infrastructure for diagnosis in medical genetics contributed to the interpretation of novel findings in future decades.

High-resolution cytogenetics assisted in the localization of multiple disorders in the human genome, but even in the postnatal context, its clinical value was limited. As in Prader-Willi syndrome, high-resolution analysis was powerful enough to aid human cytogeneticists in identifying a common mutation in many patients, but they did not regard it as sufficiently reliable to diagnose all individuals. Acknowledging this shortcoming, Ledbetter later explained, "[High-resolution cytogenetics] was always technically difficult and had a subjective component to it. . . . We need much more robust, sensitive, and objective technologies. The theme of my research interests has been improving the technology to get rid of the subjectivity and to increase the diagnostic sensitivity and accuracy of cytogenetics testing."[75] Ledbetter's expertise in deriving clinically meaningful results from this "difficult" technique benefited him significantly as he built his career around studying Prader-Willi syndrome and other microdeletion disorders during the 1980s. Eventually, however, he began to critique the "subjective component" of high-resolution cytogenetics and argued for new approaches. What human cytogeneticists had presented as an acceptable form of subjectivity in the visual interpretation of chromosomes was increasingly viewed in the late 1980s as unmanageable uncertainty and as an impediment to clinical objectives, such as expanding prenatal diagnosis.

With the increasing confusion over the role of 15q11–13 microdeletions in causing Prader-Willi syndrome, medical geneticists looked to molecular techniques as a new, more "objective" approach. Various molecular means for probing and characterizing genetic mutations promised to circumvent the subjectivity of visual interpretation that was inherent in human cytogenetic analysis. Nevertheless, a look back to the late 1970s and early 1980s reveals similar predictions of the promise for new high-resolution cytogenetic techniques to remove the problematic subjectivity inherent in clinical diagnosis. As McKusick put it in 1978, distinct chromosomal abnormalities allowed medical geneticists to isolate clinical disorders in "pure culture" so that their distinguishing features could be more precisely examined.[76] New genetic techniques often came with an allure of objectivity that eventually wore off with real-world experience, undermining the apparent rigor of any diagnostic approach and exposing subjectivities that researchers came to define as unacceptable.[77]

In the case of Prader-Willi syndrome, compounded experience by the late 1980s had demonstrated that the use of cytogenetic microdeletions to delineate disorders had its limitations. Medical geneticists were thus pushed to look for another new, seemingly more objective, diagnostic regime in emerging molecular techniques. While the introduction of molecular techniques in human cytogenetics, and medical genetics more broadly, during the 1980s and 1990s did help to clarify certain points of uncertainty in clinical diagnostics, they too failed to become the objective panacea that some had been expecting.

Chapter 4

Seeing with Molecules

Victor McKusick, in his extensive writing on the morbid anatomy of the human genome, lauded restriction enzymes, a new tool developed by molecular biologists in the 1970s. He considered them "a valuable scalpel for dissecting the human genome" and noted that the adoption of restriction enzymes by human gene mappers had contributed to the "fine structure mapping of chromosomes" at the molecular level, revealing the genomic location of hemoglobin genes on chromosome 11, among many other examples.[1] These techniques, McKusick suggested, would also allow medical geneticists to conduct higher-resolution analysis of the anatomy of the human chromosomes, which was, "in terms of resolution, . . . analogous to microscopic anatomy; by comparison, delineation by other techniques [was] analogous to gross anatomy."[2] New molecular techniques, he argued, were as revolutionary to the analysis of the human genome as the microscope had been to the clinical examination of the human anatomy. With these references, which he published in a variety of scientific and medical journals during the early 1980s, McKusick sought to bring the new tools of molecular biology to clinical attention and highlight their potential broad value in medical genetics for diagnosing and preventing disease.

Paul Berg, a molecular biologist and early participant in restriction enzyme research, argued similarly in his 1980 Nobel Prize address: "Just as our present knowledge and practice of medicine relies on a sophisticated knowledge of human anatomy, physiology, and biochemistry, so will dealing with disease in the future demand a detailed understanding of the molecular anatomy, physiology, and biochemistry of the human genome. There is no doubt that the development and application of recombinant DNA techniques has put us at the threshold of new forms of medicine." Berg shared the Nobel Prize in chemistry that year with Walter Gilbert and Frederick Sanger for their contributions to recombinant DNA, a set of techniques in which molecular biologists used restriction enzymes to cut and paste DNA into new and useful arrangements. Highlighting the immediate potential for clinical application, Berg noted that the promise of recombinant DNA in medicine would require physicians to become "as conversant with the anatomy and physiology of chromosomes and genes" as they already were with the gross anatomy of the body.[3] Molecular biologists alone, Berg claimed, could not bring these promises to fruition. Medical expertise was necessary to bridge the gap between the readily visible and the submicroscopic features of the human anatomy and to characterize the role of each in disease.

During the 1980s, as medical geneticists puzzled over the genetic etiologies of fragile X, Prader-Willi, and Angelman syndromes, they also pondered this divide: how could the unseen molecular DNA mutations that caused these disorders be made into visible and diagnostically useful markers? Human cytogeneticists believed that, even under the most favorable conditions, a chromosomal deletion needed to be millions of nucleotides long to be recognizable under the microscope. This meant that all but the largest disease-causing mutations were likely to be missed, even with the highest-resolution of human cytogenetics techniques. Mindful of this, cytogeneticists, following the enthusiastic recommendations of McKusick, began to seek out new ways of molecularizing human cytogenetics to improve clinical diagnosis.

At the same time, molecular biologists, who up to this point had largely worked in separate spaces from physicians and human cytogeneticists, followed Berg's call and increasingly moved into clinical research. Molecular biologists were drawn to begin studying clinical disorders in part by the intriguing and unexplained findings that had been emerging from cytogenetic studies of various disorders.[4] How could Prader-Willi and Angelman syndrome be caused by the same chromosomal deletion? What accounted for normal transmitting males in fragile X families? Why did specific chromosomal translocations cause cancer? In the clinical context, answers to these questions promised to lead to improvements in diagnosis, treatment, and prevention. For basic molecular biologists, the studies of genetic disorders represented new and unpredictable experimental systems that offered the possibility of uncovering novel biological phenomena.[5]

Historian Peter Keating and sociologist Alberto Cambrosio have insightfully argued that the molecular tools that entered clinical research during this period "proved to be less an application of biology to medicine than a biological solution to a problem raised in the clinic." In this and other work examining the hybrid nature of late-twentieth-century biomedicine, Keating and Cambrosio demonstrated that the introduction of molecular biology into clinical research did not reflect the unidirectional application of biological knowledge in the medical realm, resulting in the "reduction of pathology to biology," but rather the uptake of new tools to examine questions raised in the clinic.[6] Along similar lines, this chapter examines the entry of practitioners and tools from molecular biology into medical genetics during the 1980s. While this molecular influx brought new approaches, visual traces, and ways of thinking to medical genetics, molecular perspectives did not replace the centrality of chromosome-level thinking, analysis, and organization in the field. Instead, the introduction of molecular tools into human cytogenetics led to the development of a new hybrid area of research and clinical analysis within medical genetics; by the 1990s it came to be known as "molecular cytogenetics."[7]

The human cytogenetics techniques examined in preceding chapters involved the direct visualization of chromosomes under the microscope, using various procedures and stains to make them identifiable. Chapter 3 describes the desire of cytogeneticists "to bridge the gap" between chromosomes and the molecular level of DNA nucleotides.[8] Strictly speaking, however, medical geneticists understood these two levels of genetic analysis to be incommensurable. Cytogeneticists conceptualized chromosomes as three-dimensional objects comprising tightly compacted strands of DNA, whereas molecular biologists characterized and examined DNA as a one-dimensional string of nucleotides.[9] While 15q12 was a very specific chromosomal location for human cytogeneticists, molecular biologists regarded 15q12 as a loosely defined region with unknowable boundaries, spanning millions of nucleotides. Even with these insurmountable differences in scale, however, medical geneticists remained focused on localizing molecular genes and causes of disease within the existing chromosomal infrastructure. This was the case because banded chromosomal ideograms were a common point of reference for identifying genomic locations; most medical geneticists were familiar with them.

The molecularization of human cytogenetics occurred by way of the multidecade development and refinement of in situ hybridization. In the 1970s and 1980s, molecular biologists and human cytogeneticists collaborated to create new approaches for turning specific DNA sequences into microscopically visible "probes." This chapter examines the adoption of new molecular techniques and considers the challenges that human cytogeneticists faced in developing useful probes for in situ hybridization. To probe for the chromosomal location of a particular gene, cytogeneticists had to already possess a segment of DNA from it. Isolating these gene-specific DNA fragments was a difficult and time-consuming process.

To explore how in situ hybridization came to medical genetics, it is necessary first to thoroughly examine the adoption of a parallel approach: in vitro hybridization. That was the technique medical geneticists used to create some of the first molecular genetic diagnostics, which were based on restriction enzyme analysis. Importantly for this history, the identification of disease-specific restriction enzyme markers directly facilitated the isolation of DNA segments for use as in situ hybridization probes along chromosomes. Cytogeneticists struggled for years to make these probes reliably visible under the microscope. In the closing decades of the twentieth century, however, they used in situ hybridization to map many disorders within the genome's morbid anatomy and medical genetics's growing chromosomal infrastructure for diagnosis.

A Molecular Approach to Medical Genetics

Molecular biology entered medical genetics during the late 1970s and 1980s through the development and refinement of DNA hybridization techniques. While medical geneticists believed that most DNA segments of clinical interest were much too small to be made visible microscopically, they attempted to use hybridization to test for the presence or absence of DNA segments in individuals and to track how this related to clinical disease status. Hybridization relied on the specificity with which complementary strands of DNA annealed to each other. Molecular biologists had demonstrated that in DNA, adenosine (A) nucleotides were attracted to thymidine (T) and cytosine (C) to guanine (G). These chemical attractions, molecular biologists thought, would be amplified in strands of DNA by nucleotide order. Thus, they expected that a string of 1,000 single-stranded DNA nucleotides, in the presence of millions of other single-stranded DNA molecules, would preferentially hybridize to the strand that most closely corresponded to its complementary sequence. So the nucleotide strand ATCGATCGATCG, molecular biologists believed, was most likely to hybridize with the sequence TAGCTAGCTAGC if it was present. Throughout the 1960s, molecular biologists experimented with the hybridization of complementary nucleotide strands made up of DNA and RNA (a related nucleotide molecule).[10]

At the same time, cytogeneticists adopted hybridization as an approach for determining the location of specific DNA sequences along the chromosomes of various species. In 1969 Yale University cytogeneticists Mary-Lou Pardue and Joseph G. Gall proposed a new approach for using DNA hybridization to identify the location of certain categories of nucleotide sequences within a genome. Their in situ hybridization technique involved hybridizing single-stranded DNA with denatured chromosomes prepared for cytogenetic investigation. Pardue and Gall added a radioactively labeled version of the nucleotide thymidine to the tissue culture of *Xenopus* toad cells. This thymidine, which contained a radioactive hydrogen atom (^{3}H tritium), was integrated into the DNA of the cells as they replicated.[11] The researchers then allowed the radiolabeled DNA to selectively hybridize with *Xenopus* chromosomes on microscope slides. The DNA sequences that had hybridized to the chromosomes were individually too small to be resolved under the microscope. Over the coming weeks, however, the ^{3}H thymidine nucleotides within them released enough radioactivity to produce a signal that Pardue and Gall could make visible along the chromosomes, revealing the locations where the DNA had hybridized.[12]

Pardue and Gall, in their initial work, were not seeking to locate a single genetic entity. To identify the chromosomal locus of a specific gene using in situ hybridization, cytogeneticists would need a complementary nucleotide sequence. One option for obtaining one was to isolate the gene's complementary RNA transcript. Pardue and Gall demonstrated the feasibility of doing so in 1970 by using ribosomal RNA (rRNA) from *Xenopus* to identify the locations of DNA sequences coding for rRNA along the chromosomes of *Drosophila* flies and other species.[13] Cytogeneticists also sought to acquire messenger RNAs (mRNA) to use for in situ hybridization. They had few options for isolating the mRNA from most individual genes at this time; however, some clever techniques for acquiring particular mRNA sequences were developed.

Human cytogeneticist Kurt Hirschhorn and his colleagues at Mount Sinai School of Medicine in New York City extracted mRNA from red blood cells. These cells were chosen because they lacked nuclei, and thus genomic DNA, but did contain mRNAs deriving almost exclusively from the globin genes, which produce blood proteins like hemoglobin. Hirschhorn incorporated radioactive nucleotides into this mRNA and hybridized it with denatured human chromosome preparations. In a 1972 *Nature* article on this work, Hirschhorn and colleagues suggested that their technique was the first targeted approach for mapping human genes: "Until now the assignment of various genes to specific human chromosomes has been approached indirectly, by the use of marker chromosomes, deletion mapping, and cell hybridization. This has made it possible to speculate on the locus of a few random genes on the human chromosomes, where events have been favorable to such studies. But now we have a method by which gene loci can be assigned to specific areas of human chromosomes provided only that it is possible to purify mRNA specific for the locus sought."[14] This was a significant "if." While Hirschhorn's laboratory had been able to isolate globin mRNAs from red blood cells, most RNAs did not already exist in near isolation in cell lines. These geneticists sought to distinguish their approach from somatic cell hybridization, but their demonstration also relied on a "favorable event" specific to globin mRNA.

Over the following years, Hirschhorn's team and other research groups used additional, relatively easy-to-isolate, RNAs for in situ hybridization. Those RNAs were produced in great quantities by genes that existed in multiple copies, sometimes hundreds, across the human genome. In various papers during the early 1970s, Hirschhorn and colleagues presented in situ hybridization as "potentially the most powerful technique available for gene mapping."[15] Despite the long-term promise of this technique, however, cytogeneticists knew that to map most human genes, they would first need a more direct approach for isolating gene-specific nucleotide sequences for in situ hybridization. For this, they turned to the emerging tools of molecular biology during the 1970s and 1980s. The next sections describe how the integration of molecular biology into medical genetics was broader than efforts to improve in situ hybridization. Medical geneticists adopted various molecular tools for diagnosing genetic disorders. Importantly, these studies eventually contributed to the development of molecular approaches in cytogenetics as well.

In vitro Hybridization

Molecular biology also entered medical genetics during the 1970s through parallel in vitro approaches. Rather than in situ hybridization along chromosomes in cell preparations, medical geneticists conducted in vitro hybridization outside of cells in test tubes, gels, and filters. The aim of that technique was quite similar: medical geneticists sought to determine the relative presence or absence of a genomic segment in different individuals and compare this to clinical disease status. In the mid-1970s, medical geneticists Yuet Wai Kan, Michael Golbus, and Andrée Dozy at the University of California, San Francisco, adopted in situ hybridization so that they could offer prenatal testing to a Chinese-American couple who had lost two late-term pregnancies to inherited α-thalassemia. Kan and Golbus had previously provided prenatal diagnosis for β-thalassemia in other families by visually examining fetal blood cells. But sampling fetal blood was a difficult and rarely performed technique that carried a significant risk of miscarriage. Mindful of these challenges, the clinicians decided that a prenatal test relying on the hybridization of DNA collected from amniotic fluid would be preferable. Compared to fetal blood sampling, amniocentesis was an easier and safer procedure for prenatal diagnosis that was increasingly being used for cytogenetic testing to identify Down syndrome and other disorders.[16]

Kan, Golbus, and Dozy published a report in the *New England Journal of Medicine* in 1976 demonstrating that in vitro hybridization techniques could be used to characterize the number of α-globin genes present in a patient. In general, people with three or more α-globin genes appeared clinically normal, while those with one or two were affected by anemia resulting from α-thalassemia. Fetuses with zero active copies of this gene died in late pregnancy or soon after birth. Kan, Golbus, and Dozy examined DNA samples from each family member, as well as the fetus, by hybridizing them with radiolabeled α-globin cDNA in vitro. Next, they assessed the number of genes present in each individual based on the person's clinical status and the relative amount of hybridization that had occurred. The clinicians found that the fetus had a reduced but detectable presence of α-globin DNA, suggesting that it would be affected by α-thalassemia but did not have a lethal form. Using this molecular diagnostic approach, medical geneticists were able to assess α-thalassemia's likely impact much earlier in its life history than was previously possible, offering the chance for prevention by abortion while posing little risk to the fetus.[17]

In an editorial accompanying Kan, Golbus, and Dozy's report, David G. Nathan, a pediatrician and hematologist at Harvard Medical School, hailed the success of their approach for prenatal diagnosis of α-thalassemia, while

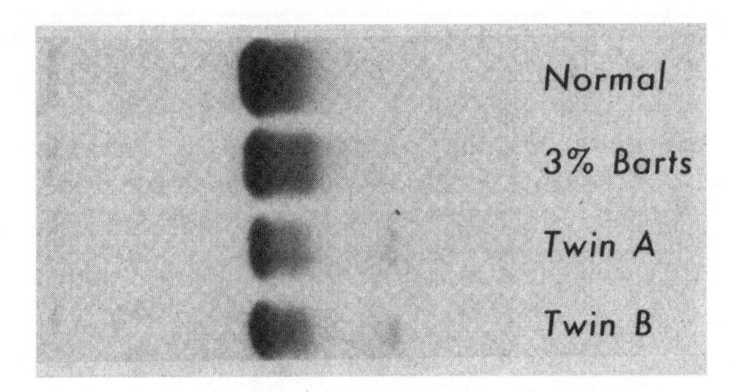

Gel electrophoresis demonstrating the relative number of α-globin genes present in differing forms of α-thalassemia. *Source:* Y. W. Kan, M. S. Golbus, and A. M. Dozy, "Prenatal Diagnosis of Alpha-Thalassemia: Clinical Application of Molecular Hybridization," *New England Journal of Medicine* 295, no. 21 (1976): 1166. Copyright © Massachusetts Medical Society. Reprinted with permission from Massachusetts Medical Society.

noting that it was still a technically difficult test to perform and had a high likelihood for error. Nathan suggested that the extraction of five times the amount of fetal DNA used by those researchers would improve the procedure's reliability by helping to overcome any "background" signal from hybridization, which might be incorrectly interpreted as representing the presence of one α-globin gene in a fetus that actually had zero.[18] Collecting significantly more DNA was a particular challenge in the prenatal setting, though, because it required culturing amniotic fluid cells for a longer period of time. Amniocentesis in this case was performed at 15 weeks and was followed by 5 weeks of culturing. Extra time for producing more DNA in cell culture would push diagnosis later into the second trimester, which physicians feared would make a potential abortion more difficult.

As prenatal diagnosis was becoming more commonplace, the timing of preventive abortions was a significant concern of obstetricians and medical geneticists. In response, some worked to develop new sampling techniques that could be performed earlier in pregnancy and used to generate genetic results more quickly. One eventual alternative to amniocentesis was chorionic villus sampling (CVS), a procedure that was performed during the first trimester. While this procedure remained experimental, risky, and rare until the late 1980s, it allowed for earlier testing and did not require weeks of tissue culturing for genetic results. Along with new molecular diagnostic techniques, further described in the next section, many medical geneticists pointed to CVS as a significant innovation for improved prenatal diagnosis and prevention.[19]

Restriction Enzymes in the Clinic

During the 1980s, restriction enzyme analysis also brought molecular biology into clinical practice. A vocal proponent of restriction enzymes' expanding uptake in medicine, Victor McKusick highlighted their clinical value in many of his publications that presented genetics to physicians.[20] In the 1970s molecular biologists had purified various restriction enzymes from different species of bacteria and had found that each cut DNA at a specific nucleotide sequence.[21] In subsequent years, molecular biologists at Stanford University and at the University of California, San Francisco, developed methods for pasting the resulting DNA fragments together in new and useful ways. This process and its application in molecular genetics came to be known as recombinant DNA.[22]

In 1978 Kan and Dozy adopted restriction enzymes with the goal of improving their prenatal diagnostic assays for inherited β-globin disorders, beginning with sickle-cell anemia. Using a restriction enzyme, Kan and Dozy cut DNA samples from various members of an affected family into fragments. Molecular biologists believed that variants, or "polymorphisms," in the length of DNA fragments cut by restriction enzymes were quite common among different individuals, because of inherited differences in the DNA sequence at particular cut sites. For instance, a restriction enzyme might make a cut at a certain location near the β-globin gene in some individuals but not others because of an inherited polymorphic change in their DNA code. This difference in DNA sequence likely had nothing to do with causing sickle-cell anemia, but it might have been inherited along with, or "linked" to, a sickle-cell mutation, thus facilitating diagnosis.

To make cut site variations visible within a family and to associate the differences with each person's clinical status, Kan and Dozy used their existing β-globin gene "probe" for in vitro hybridization analysis. DNA fragments from the family members were placed in an agarose gel and separated by size, using an electric current. With this process, called gel electrophoresis, Kan and Dozy showed that the segment of DNA they probed using radiolabeled mRNA was cut into two differently sized fragments among the family members, corresponding with their clinical

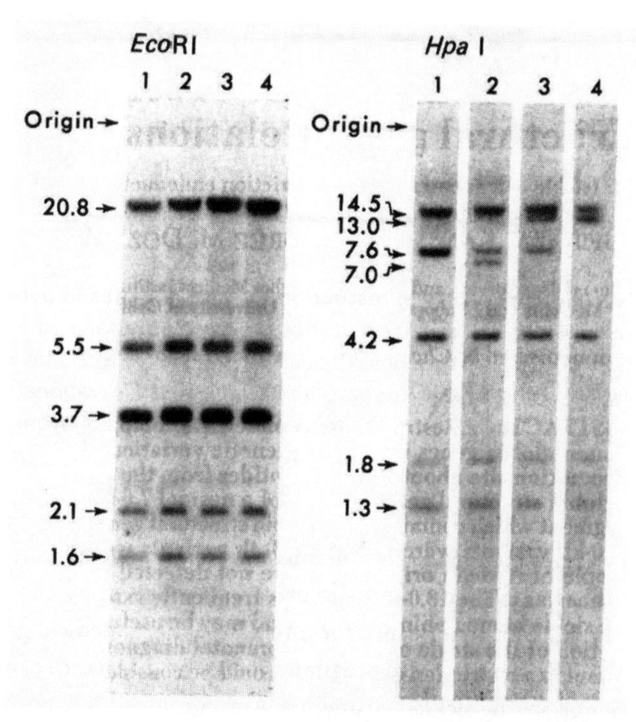

Gel electrophoresis demonstrating outcomes from the restriction enzyme digest analysis of the β-globin gene. Patients who were sickle-cell carriers showed bands at 7.6kb and 13.0kb. Those who were affected by the disorder had a band only at 13.0kb. *Source:* Y. W. Kan and A. M. Dozy, "Polymorphism of DNA Sequence Adjacent to Human Beta-Globin Structural Gene: Relationship to Sickle Mutation," *Proceedings of the National Academy of Sciences* 75, no. 11 (1978): 5632. Reprinted with permission from Yuet Wei Kan.

sickle-cell anemia status. The researchers noted that a DNA fragment estimated to be 7,600 nucleotides in length (7.6kb) was associated with a normal β-globin gene, while a 13.0kb fragment was found in individuals with a mutated gene. The gene mutation appeared to be closely linked to a polymorphism that disrupted a *Hpa*I cut site. The sequence change was present on chromosomes with a normal copy of the β-globin gene but absent on those with a mutant copy.[23]

Kan and Dozy published a report that same year in *Lancet* on the successful use of their restriction enzyme technique to provide a sickle-cell anemia diagnosis prenatally. The family studied had one parent who was a sickle-cell trait carrier and one who had two mutant copies of the β-globin

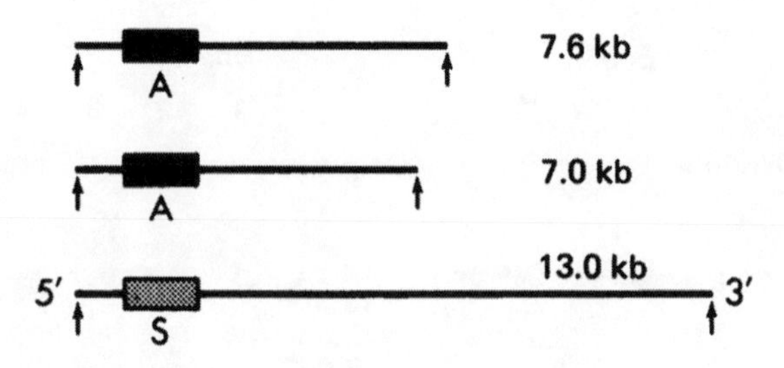

Diagram showing the differing restriction enzyme cut sites in patients with and without a sickle-cell disease mutation. *Source:* Y. W. Kan and A. M. Dozy, "Polymorphism of DNA Sequence Adjacent to Human Beta-Globin Structural Gene: Relationship to Sickle Mutation," *Proceedings of the National Academy of Sciences* 75, no. 11 (1978): 5634. Reprinted with permission from Yuet Wei Kan.

gene and was affected by sickle-cell anemia. This couple already had two children; one was a carrier and the other was affected by the disorder. Based on Mendelian inheritance for a recessive trait, medical geneticists believed that the couple's third pregnancy had a 50-50 chance of either being a carrier or having sickle-cell anemia. Amniocentesis was performed at 18 weeks. Unlike the prenatal test that Kan and Dozy reported in 1976, this test, with sensitivity provided by in vitro hybridization following treatment with a restriction enzyme, allowed diagnostic analysis to be performed on the small amount of DNA already present in the amniotic fluid sample. This meant bypassing multiple weeks of cell-culturing time necessary in previous prenatal diagnostic approaches. The difference in timing was of significant value in the prenatal context, offering an earlier answer for the anxious couple and moving the option of preventive abortion earlier in the second trimester. Ultimately the fetus was found to have one 7.6kb restriction fragment and one that was 13.0kb, suggesting that it was a sickle-cell trait carrier.[24]

Kan and Dozy's report presented a model for molecular diagnosis using in vitro hybridization that could hypothetically be extended to any inherited genetic disorder. Instead of relying on the rare instance of a visible chromosomal abnormality, their approach used a nonpathogenic restriction fragment variant, something that medical geneticists believed to exist in great quantity and with sufficient variability in all individuals' genomes. Also importantly, to apply this diagnostic technique in another condition, medical geneticists did not need a complementary probe that targeted its causative mutation. All they required was a polymorphic restriction enzyme cut site that was closely linked to the disorder's molecular etiology, and a probe that would hybridize somewhere in this DNA segment, so it could be made visible on a gel.

Notably, this molecular approach to prenatal diagnosis involved the production of a set of visual traces that were new to the scientific and medical literature of the late 1970s but were becoming increasingly prevalent in it. The interpretation of these images involved a logic that differed from that associated with cytogenetics, where chromosomes were directly used as a reference for assessing the location of a gene or a mutation. With restriction enzyme analysis, the common point of reference was instead the relative length of the DNA fragments that were produced and probed. Rather than looking for "pathological" chromosomes, in comparison to standardized conceptions of "normal" ones, geneticists used restriction enzyme analysis to compare individuals, and identify DNA sequence variants, with no established standard for reference. While in cytogenetics the absence of a chromosomal band directly indicated the deletion of a certain genomic segment, in molecular genetic analysis the absence of a band of DNA following gel electrophoresis suggested a cut site polymorphism. Medical geneticists used this finding to differentiate people, but it was not interpreted as implying pathology on its own. Restriction enzyme analysis brought to medical genetics a new visual logic, which became increasingly common after 1980. This molecular way of seeing complemented but did not supplant chromosomal ways of seeing and organizing knowledge in medical genetics.[25]

Libraries of Molecular Landmarks

While medical geneticists developed new molecular diagnostics in the late 1970s using in vitro hybridization, human cytogeneticists remained focused on identifying in situ hybridization probes for chromosomal disease mapping and diagnosis. Any piece of DNA that was isolated from the human genome could be used as a probe for hybridization analysis. However, cytogeneticists knew that not all segments of genomic DNA were created equal. In the late 1960s, molecular biologists had found that much of the DNA in plant and animal genomes was repetitive and did not code for genes. The repetitive DNA came to be known during the 1970s by the derisive term "junk DNA."[26] While segments of repetitive DNA hybridized with many different locations in the genome, probes containing unique sequences from single genes or their surrounding sequences were expected to hybridize selectively to only one or a few sites. These were the probes that human cytogeneticists sought for mapping genes and genetic disorders.

During the late 1970s, molecular biologist Tom Maniatis and his colleagues at the California Institute of Technology began creating "libraries" of genomic segments from humans and other animals. Geneticists hoped that the libraries would be a resource for isolating nonrepetitive genomic sequences. Maniatis and colleagues used restriction enzymes to cut genomic DNA into many segments, 15–20kb in size, and then inserted these fragments into rings of DNA called plasmids, which were copied and stored in bacteria. In one day of bacterial replication, millions of copies of each segment were produced within a separate bacterial colony.[27] Molecular biologists then screened the colonies with radiolabeled probes to determine whether they contained a genetic fragment with a repetitive or a unique *DNA* sequence.[28] Those identified as having a unique sequence were grown further and stored indefinitely in a freezer for later use or distribution to other researchers. Over the following years, many researchers were given access to this library.

Other researchers developed more targeted libraries containing DNA from individual human chromosomes, which were isolated using somatic cell hybridization techniques or a newly developed approach called fluorescence activated cell sorting. Physicist Joe W. Gray and colleagues at Lawrence Livermore National Laboratory in California found that many human chromosomes, when stained with a fluorescent dye, produced a distinct enough signal to be distinguished and separated out at a high rate using a machine called a flow cytometer.[29] This facilitated the isolation of large numbers of individual human chromosomes, which were digested into fragments with restriction enzymes and used to construct chromosome-specific DNA libraries. These were particularly useful for producing probes for disorders that were already associated with a certain chromosome, such as fragile X and Prader-Willi syndrome.[30]

While medical geneticists were primarily interested in finding DNA segments that were linked to genes and disease-causing mutations in genomic libraries, molecular biologists also put these libraries to use for identifying variants in restriction enzyme cut sites and the DNA fragments they produced, among different people. In an influential 1980 paper, MIT molecular biologist David Botstein suggested that identifying about 150 of these restriction fragment length polymorphisms (RFLPs) across the human genome would be an important step toward mapping many more disorders and genes to particular loci. Although most RFLPs (pronounced "riff-lips") would be outside of genes, many would be closely linked to them. Mapping a couple hundred RFLPs spread evenly across the human genome, he argued, would provide a dense set of molecular markers, at least one of which would be closely linked to every gene. As scholars Peter Keating and Alberto Cambrosio have incisively noted, Kan and Dozy had already identified a clinically useful RFLP two years earlier, which was linked to the β-globin gene.[31]

Human gene mappers located genes in the 1970s by correlating them with known physical locations in the genome, most prominently the standardized infrastructure of chromosomal bands. Another way to locate a gene was through linkage analysis to another measurable genetic variant. Described in chapter 2, this approach was used in somatic cell hybridization. If two genetic traits were usually inherited together, geneticists inferred that they were present in close proximity on the same chromosome. This technique was limited, however, by the relative lack of variability in the expression of genes located at known genomic locations. RFLPs offered a new set of landmarks that Botstein and colleagues believed were likely to be highly variable in the human population since they rarely had a phenotypic impact.[32]

In a 1980 editorial published alongside Botstein's paper, physician David Comings referred to RFLPs and other new recombinant DNA techniques that were under development around 1980 as being reflective of the "new genetics," a term that had been in use for decades to describe molecular genetics.[33] While the term was not new, there was a strong argument to be made that the approaches facilitated by restriction enzymes were quite novel. In Comings's words,

> One of the major goals when studying specific genetic diseases is to find the primary gene product, which in turn leads to a better understanding of the biochemical basis of the disorder. The bottom line reads, "This may lead to effective prenatal diagnosis and eventual eradication of the disease." But we now have the ironic situation of being able to jump right to the bottom line

> without reading the rest of the page, that is, without needing to identify the primary gene product or the basic biochemical mechanism of the disease. The technical capability of doing this is now available. Since the degree of departure from our previous approaches and the potential for this procedure is great, one will not be guilty of hyperbole in calling it the "New Genetics."[34]

Comings highlighted Kan and Dozy's approach for determining sickle-cell anemia status as an exemplar for the medical genetics to come, which would involve the diagnosis of disorders even in the absence of direct access to, or biochemical understanding of, causative genes. In the short term, this approach offered the chance to "jump right to the bottom line" and immediately provide options for prenatal diagnosis and prevention. Looking ahead, medical geneticists also believed that RFLPs would help to advance their larger, systematic goal of isolating disease-causing genes and locating them in the genome's morbid anatomy using in situ hybridization.[35]

RFLPs in Clinical Diagnosis

Following Kan and Dozy's model, in the early 1980s clinicians and molecular biologists worked collaboratively to identify RFLPs that were linked to common and devastating inherited disorders, including cystic fibrosis and Huntington's disease, which could not be diagnosed prenatally or presymptomatically. At the time, neither disorder was mapped to a specific human chromosome. A team searching for the genetic basis of Huntington's disease was led by James Gusella, a postdoctoral fellow working in the molecular biology laboratory of David Housman at MIT, and clinical psychologist Nancy Wexler. As part of her research, Wexler spent three months in Venezuela, creating an extensive pedigree of a large extended family that was affected by Huntington's disease and collecting blood samples from each living individual.

Huntington's disease, which caused severe muscular, mental, and behavioral decline beginning in early middle age, had been regarded for decades as having a Mendelian dominant inheritance pattern. Since clinical signs of the disorder often were not apparent until after age 40, most people had children before finding out they were affected. This was particularly troubling because those who were affected had a 50% chance of passing the disorder on to each of their children.

Wexler's extensive pedigrees of Huntington's families offered the opportunity to look for linkages between affected individuals and specific RFLPs. Using the blood samples provided by Wexler, Gusella began searching for a RFLP marker that was closely associated with the Huntington's gene. He drew upon DNA fragments from Tom Maniatis's genomic sequence library. Gusella believed that it would likely take years of work, testing thousands of probes, to find one that was closely linked to Huntington's disease. In a miraculous stroke of luck, though, he detected linkage on just the twelfth probe he tested. Wexler later used a geographic metaphor to illustrate the vastness of the space in which this tiny marker had been blindly sought: "It was as though, without the map of the United States, we had looked for the killer by chance in Red Lodge, Montana, and found the neighborhood where he was living."[36]

Gusella had been looking for a RFLP that produced variant DNA fragments that correlated with the clinical presence or absence of Huntington's disease in middle-aged and older individuals. One of the probes, called G8, contained a RFLP for the restriction enzyme *Hind* III, which produced numerous fragments that varied among individuals. G8 was considered an "anonymous" probe, because it was not associated with a gene sequence or chromosomal location.[37] Gusella had also been working with a pedigree from a Huntington's family in Iowa that initially suggested linkage between the G8 probe and the disorder. Ultimately, however, it was the large pedigree collected by Wexler that provided sufficient data to demonstrate a complex but strong pattern of linkage between G8 RFLPs and the Huntington's disease gene.

After having found one of the Huntington's disease gene's closest RFLP neighbors, Gusella and colleagues next sought to determine at what chromosomal address the gene was located. For this, they turned to a panel of somatic cell hybrid lines. Using restriction enzyme analysis, they found that the in vitro hybridization pattern of the G8 probe matched most closely with other probes known to be located on chromosome 4.[38] A more exact location for the Huntington's disease gene on chromosome 4 was not determined. Nevertheless, associating it with a specific RFLP marker and genomic location was regarded by medical geneticists as a large step forward in understanding the disorder and, potentially, in diagnosing it presymptomatically.

Gusella and colleagues pointed to the use of RFLP analysis in this case as a model for identifying the genetic location of additional single-gene disorders, claiming in 1983, "This study demonstrates the power of using linkage to DNA polymorphisms to approach genetic diseases for which other avenues of investigation have proved unsuccessful. It is likely that Huntington's disease is only the first of many hereditary autosomal diseases for which a DNA marker will provide the initial indication of chromosomal location of the genetic defect."[39] During the mid-1980s, multiple research groups adopted similar molecular approaches to map the genomic location of additional disorders.

In another notable early instance of success with RFLP analysis, several papers were simultaneously published in *Nature* and *Science* in November 1985 reporting the identification of multiple genetic markers that were linked to cystic fibrosis. This disorder caused the buildup of thick mucus in the lungs and often led to death by age 40. It was regarded as following a Mendelian recessive inheritance pattern. Various research groups, with members in the United States, Canada, the United Kingdom, France, and Denmark, worked for many years to test more than 100 RFLP probes for linkage to cystic fibrosis in affected families. Their efforts eventually led to the identification of a few closely linked RFLPs. One of these markers was estimated to be 15 million base pairs from the cystic fibrosis locus. This was still a significant distance, but it did exclude 99% of the human genome from the search for the gene. Additional markers linked to the disorder were used to locate its genetic cause on chromosome 7, providing the potential for diagnosis in at-risk families. In identifying RFLP markers for these disorders, medical geneticists sought to reshape their life histories by offering presymptomatic diagnosis and the option for prevention via prenatal detection and abortion.[40]

RFLP analysis was also applied to improving the diagnosis of fragile X syndrome. Because of the disorder's association with a visible chromosomal abnormality at Xq27, geneticists began their search for useful probes using X chromosome–specific libraries of DNA fragments. Researchers were looking for RFLPs that could be used to distinguish the "fragile" copy of the X chromosome from the normal copy in female carriers. A closely linked RFLP could aid in the identification of carriers, as well as affected fetuses, in whom the fragile X site's cytogenetic expression was variable and thus not sufficiently reliable for prenatal diagnosis.

There were already many molecular markers available in the fragile X region of the genome—unlike the cases of Huntington's disease and cystic fibrosis. Indeed, many other human genes had been located in this area owing to their X-linked inheritance pattern. But molecular markers that could provide reliable prenatal diagnosis remained largely out of reach. In one instance, reported in 1988, a fetus estimated, by RFLP analysis, to have a 98% chance of having fragile X syndrome developed normally after birth, demonstrating that the prenatal molecular diagnosis had been a false positive result. RFLP studies of the fragile X site were also aimed at finding a causative gene for the disorder, assuming such a gene existed.[41] Seven years of work went into the isolation of the fragile X gene in 1991, following significant international collaboration.

As discussed in chapter 1, the identification of a gene and a molecular mechanism for fragile X syndrome answered many questions about its genetic basis and facilitated the development of more efficient and reliable testing for carriers and affected fetuses. Clinical uncertainty after years of cytogenetic analysis of fragile X syndrome drew the interest of molecular biologists to this clinical conundrum. Many were rewarded for their interest, thanks to the identification of a new biological phenomenon. Molecular biologists were similarly drawn to the study of Prader-Willi and Angelman syndromes, described in chapter 3, which despite having distinct clinical features had been associated with the same cytogenetic mutation. These findings conflicted with the one mutation–one disorder ideal of genetic medicine and threatened the reliability of presymptomatic diagnostics. Many medical geneticists believed that molecular examination of the deleted region would circumvent the "subjectivity" of cytogenetic analysis and provide clarity by identifying distinct Prader-Willi and Angelman mutations at the molecular level.[42]

One Mutation–One Disorder?

During the closing years of the decade, molecular biologists in Samuel Latt's laboratory at Harvard University were actively searching for RFLPs in the Prader-Willi syndrome deletion region of chromosome 15. In 1986 postdoctoral fellow Timothy Donlon used copies of chromosome 15 from normal and affected individuals to create DNA libraries for the isolation of hybridization probes specific to the Prader-Willi region.[43] Donlon soon left Boston for a position as director of molecular and clinical cytogenetics at Stanford University Hospital. However, over the coming years, other members of the Latt laboratory used Donlon's probes to examine patients diagnosed with Prader-Willi syndrome, many of whom had normal chromosomes under the microscope but were expected to have small mutations in the 15q11–13 region.[44]

In 1989 postdoctoral fellows Robert Nicholls and Joan Knoll used RFLP analysis to demonstrate that in Prader-Willi patients the copies of chromosome 15 containing the causative 15q11–13 deletion were always inherited paternally. Fathers did not have the deletion in their own cells; rather, it came about during sperm production. Nicholls and Knoll also recognized that this correlation pointed to an explanation for the unanticipated finding two years earlier that the Prader-Willi deletion also appeared in patients with the clinically distinct disorder Angelman syndrome. They found that individuals with the two disorders had the same deletion, but those with Angelman syndrome always had the deletion on the maternally inherited copy of chromosome 15, distinguishing it from the paternally inherited Prader-Willi syndrome.[45]

Nicholls and Knoll argued that this unorthodox situation in which two distinct disorders were caused by the same deletion, or sometimes apparently no deletion at all, resulted from a phenomenon known as "imprinting," which had recently been described in mice. Imprinting involved the "epigenetic" addition of methyl groups to certain DNA nucleotides, which resulted in the silencing of certain genes. Methylation was also later found to cause fragile X syndrome. The genes on chromosome 15 that were silenced by methylation varied between the paternally and maternally inherited copies of chromosome 15. Having a deletion on the paternal copy of chromosome 15 meant that a child received only one copy of those genes, some of which were methylated on the maternally inherited chromosome and therefore completely absent in terms of clinical expression. The opposite case of a maternal deletion and methylated paternal copy of chromosome 15 caused Angelman syndrome. In addition, inheriting two maternal copies of chromosome 15 and no paternal copy, a condition called "uniparental disomy," also caused Prader-Willi syndrome and accounted for some instances in which no deletion was identified. In sum, Prader-Willi syndrome was caused by the lack of expression in paternally inherited genes, and Angelman syndrome by the lack of expression in maternally inherited genes.[46]

It was by this complicated epigenetic mechanism that a most unexpected thing happened, which seemingly complicated the one mutation–

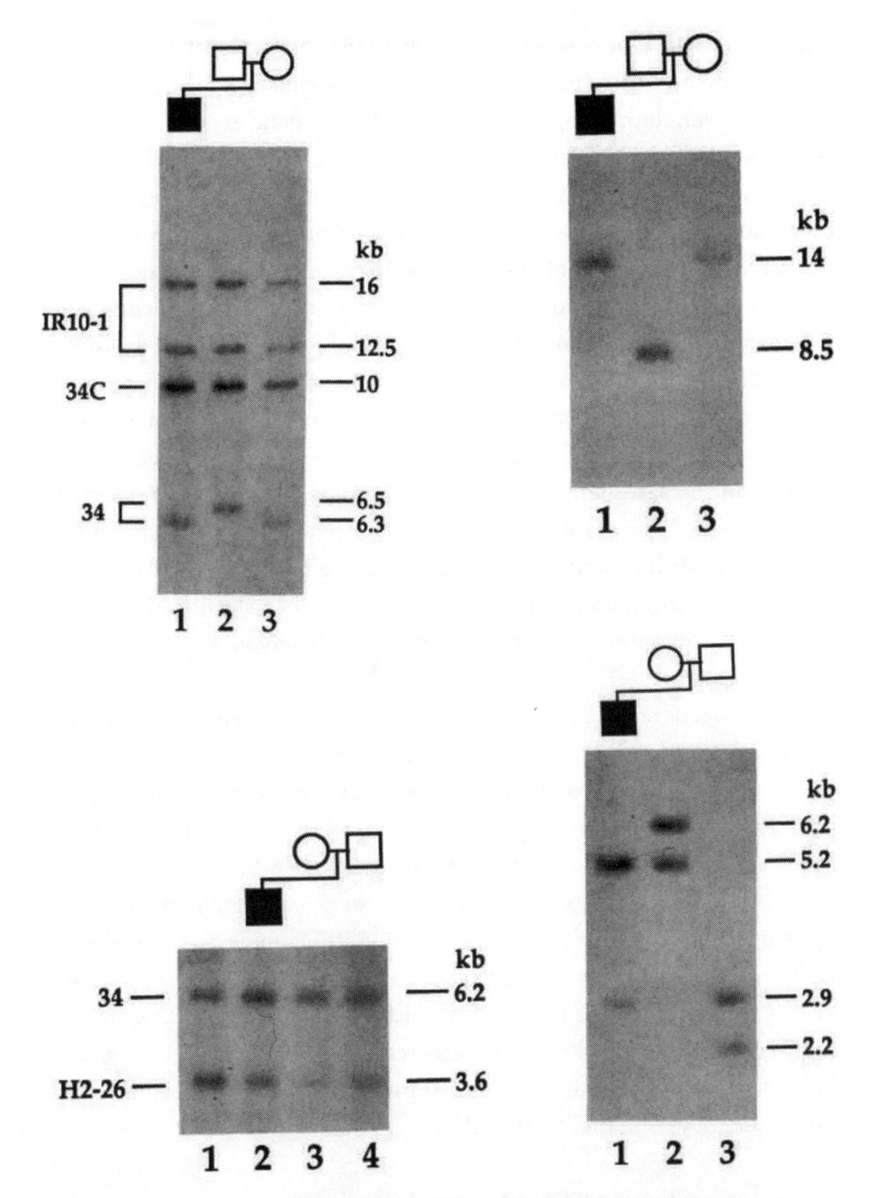

Gel electrophoresis demonstrating uniparental disomy of the maternal copy of chromosome 15 in a Prader-Willi syndrome patient. *Source:* R. D. Nicholls et al., "Genetic Imprinting Suggested by Maternal Heterodisomy in Non-Deletion Prader-Willi Syndrome," *Nature* 342, no. 6247 (1989): 283.

one disorder ideal of postwar medical genetics. At the level of DNA code, Prader-Willi and Angelman syndrome could be caused by the same mutation. However, imprinting, another dimension of gene expression that had never previously been described in humans, could make the clinical outcomes radically different. Faced with a clear contradiction, the one mutation–one disorder model was saved in this case by a significant corollary: it mattered on which copy of chromosome 15 the mutation occurred.

In regard to this history of molecular techniques entering clinical genetics, it is worthwhile to note that, while RFLP analysis played a central role in the work of Nicholls and Knoll, all of the clues that had gone into solving this puzzle had been available years earlier. Cytogeneticist Merlin Butler had demonstrated in 1983 that Prader-Willi deletions were always present on the paternally inherited chromosome. Four years later, cytogeneticist Ellen Magenis had found that Prader-Willi and Angelman syndrome patients had similar deletions under the microscope.[47] Around the same time, cytogeneticist Bruce Cattanach described the differential phenotypic outcomes resulting from uniparental disomy in mice, the basis of imprinting.

As cytogeneticist David Ledbetter later put it, "When Rob Nicholls published his paper, we all kicked ourselves for not figuring it out, because we should have been able, based on the mouse literature. If any of us had paid attention to imprinting in the mouse, we should have been able to predict this."[48] All of the pieces were already in place, based on the work of cytogeneticists. Ultimately, the common genetic basis of Prader-Willi and Angelman syndromes was determined using RFLP analysis, but molecular techniques were not necessary to generate this outcome.[49] The remainder of this chapter explains how, though molecular analysis became increasingly central to medical genetics research in the 1980s, it did not replace but rather complemented the long-standing role of chromosomal mapping and analysis in medical genetics.

Visible Flags on Chromosomal Continents

During the 1980s recombinant DNA techniques aided researchers in their search for genes and DNA mutations associated with various human disorders. Identifying one or more RFLPs that were closely linked to the apparent genetic cause of Huntington's disease and fragile X syndrome provided medical geneticists valuable footholds from which they could "go for the gene," as well as molecular landmarks that could be used in presymptomatic and prenatal diagnosis. From a human gene mapping perspective, however, identifying a linked RFLP alone did not reveal the genomic location of a disorder. To determine on which chromosome, and thus where in the genome, Huntington's disease and cystic fibrosis genes were actually located, medical geneticists turned to in situ hybridization along chromosomes. Far from upending the cytogenetic techniques developed in the 1970s, molecular approaches continued to depend on a chromosomal way of seeing, and they were increasingly integrated by cytogeneticists, leading to the development of the hybrid area of molecular cytogenetics in the 1990s.

To identify the chromosomal addresses of cystic fibrosis and Huntington's disease within the genome's morbid anatomy, cytogeneticists drew upon the same probes that molecular biologists had used to identify RFLPs linked to these disorders, and put them to a new use for in situ hybridization along chromosomes. The probes had a unique sequence, which cytogeneticists believed would selectively hybridize with one genomic site: they were flags to be planted at the location of a single gene among the chromosomal continents. With these gene-specific probes for cystic fibrosis and Huntington's disease in hand, the remaining problem that cytogeneticists faced was to make the location of their tiny flags visible under the microscope.

As historian Angela Creager has impressively documented, thanks to US government promotion, radioactive tracers were put to widespread use in postwar biology research, including human cytogenetics. For the purposes of in situ hybridization, cytogeneticists needed tracers that produced enough signal to be seen in a single copy, but not so much that they were difficult to pinpoint to a specific chromosomal address. The three most commonly used by cytogeneticists were ^{3}H, ^{32}P, and ^{125}I. Among these, ^{125}I produced so much radioactivity that it lacked specificity, and ^{3}H often did not generate a bright enough signal to be seen under the microscope in single copy, meaning that it lacked sufficient sensitivity to visualize an individual probe.[50]

During the early 1980s, researchers were working on various approaches to circumvent these shortcomings. Some of the most significant work took place at cancer research centers around the United States. Cytogeneticists had demonstrated over the previous two decades that multiple cancers were caused by chromosomal rearrangements, and they also had described characteristic patterns of chromosomal abnormalities in cancer cells. As cytogeneticist and cancer researcher Janet Rowley later stated, collaboration between human cytogeneticists and molecular biologists, what she called "molecular cytogenetics," was "a Rosetta Stone for understanding cancer." The hybridization of cytogenetics and molecular biology had facilitated new and highly productive research approaches, which Rowley argued had contributed to the prestige of both areas.[51]

At the M. D. Anderson Cancer Center in Houston, geneticist Mary E. Harper and molecular biologist Grady F. Saunders collaborated to enhance the sensitivity of in situ hybridization for cancer cytogenetics research. Their approach involved making probes that hybridized not only with chromosomal DNA, but also with each other. The goal was to elicit the formation of a network of radiolabeled probes, thus making hybridized probe locations more readily visible. One probe would be planted at a specific chromosomal site, and others would bind to it. All of the probes together would emit a localized radioactive signal, which Harper and Saunders believed would be more visible microscopically. They demonstrated the value of this method in 1981 by using it to locate the human insulin gene on the short arm of chromosome 11.[52]

At the same time, other researchers began experimenting with nonradioactive approaches. Cytogeneticists hoped that different techniques could better balance specificity and sensitivity while also circumventing some of the other challenges posed by using radioactivity, including safety and disposal issues as well as lengthy exposure times. At the Fox Chase Cancer Center in Philadelphia, cytogeneticist George T. Rudkin and colleagues developed rabbit antibodies that bound to chromosomes. Going further, the researchers incorporated goat antibodies that selectively bonded to the rabbit antibodies. These goat anti-rabbit antibodies were also linked to a fluorescent compound, which could be made visible under the microscope, revealing the locations of rabbit antibody attachment along the chromosomes.[53] This new "immunofluorescent" approach combined the molecular specificity of immune system antibodies with fluorescent tags to help make specific cellular targets visible.

In 1977, soon after the publication of the Fox Chase approach for immunofluorescent visualization of proteins, Rudkin collaborated with Tufts University biochemist B. David Stollar to develop a technique for locating sites of in situ probe hybridization among *Drosophila* giant polytene chromosomes. The researchers used ribosomal RNA as a probe, which they knew would hybridize at various locations along the chromosomes. They expected that the hybridized probes would be uniquely visible in this case because polytene chromosomes have more than 100 strands of the same genetic material, combined into one body. Rudkin and Stollar expressed confidence that the sensitivity of their approach would be sufficient to visualize "a unique gene" among the polytene chromosomes, where many copies were present in a condensed location. Making visible the location of a single gene along a human chromosome, however, with just one copy to hybridize, would require further enhancement.[54]

Additional approaches to visualizing nonradioactive probes were also in development during the late 1970s and early 1980s. Some researchers attached fluorescent microspheres and other fluorescent chemicals directly to the nucleotides making up hybridization probes.[55] Others tagged the probes with compounds that would be visible under electron microscopy, such as gold spheres and the vitamin biotin.[56] In 1982 Pennina Langer-Safer and colleagues in David C. Ward's cancer genetics laboratory at Yale University reported on a technique that combined the biotin labeling of probe nucleotides with Rudkin and Stollar's immunofluorescent approach. Rabbit antibodies that targeted biotin were used to identify the hybridized probes in situ. As before, fluorescence-labeled anti-rabbit goat antibodies were then introduced to reveal the location of the hybridized probes under the microscope. This two-step, "indirect" process was important for enhancing the signal, as multiple goat antibodies would attach to each rabbit antibody, greatly enhancing the microscopic visibility of a single probe.[57]

A significant advantage of the Yale group's technique over radiolabeling approaches to in situ hybridization was that fluorescent markers produced a visible signal under the microscope immediately, rather than after a week or more of radioactive exposure. However, in contrast with the use of radiotracers, the fluorescent signal could not be visualized directly alongside chromosome bands. This was a significant problem for cytogeneticists, who wanted to use the immunofluorescence technique to correlate probe hybridization with a specific banding address in the human chromosomal infrastructure. To circumvent this problem, researchers also used goat antibodies that were linked to peroxidase, a chemical tracer that could be seen alongside chromosomal banding patterns under the microscope.[58] During the 1980s, medical geneticists had great success in using these techniques to locate disorders in the chromosomal infrastructure.

FISHing for Gene Probes

Following the localization of Huntington's disease on chromosome 4 in 1983, a team of Dutch molecular cytogeneticists led by Mels van der Ploeg and J. E. Landegent utilized nonradiographic in situ hybridization to determine the chromosomal location of the Huntington's disease–linked G8 probe. The researchers adopted an approach similar to the one developed in Ward's laboratory, using rabbit antibodies followed by peroxidase-labeled swine anti-rabbit antibodies for detection. Based on the results of in situ hybridization, they argued that the chromosomal address of the Huntington's disease gene was 4p16.3.[59] This result differed from what Gusella's group had suggested a year earlier. They had used a different approach, adopting radioactively labeled probes to perform in situ hybridization. Gusella, who collaborated with cytogeneticist Ellen Magenis, had used multiple copies of chromosome 4 with differently sized deletions to improve upon the limited resolution of radiolabeled probing. They had identified 4p16.1 as the chromosomal address of the Huntington's gene.[60] The Dutch group's differing result, which was later verified by others, demonstrated the improved resolving power of nonradioactive techniques for in situ hybridization in the localization of single human genes.

Further development of fluorescent probes continued at Lawrence Livermore National Laboratory during the mid-1980s. Physicist Daniel Pinkel and biologist Barbara Trask worked in Joe Gray's laboratory, which had previously developed fluorescence-activated cell sorting. Gray's group contributed to the improvement of what they called "fluorescence in situ hybridization," a technique that increasingly became known by the acronym FISH during the 1990s. The group developed libraries of probes for each human chromosome and used them with in situ hybridization to fluorescently "paint" each chromosome in a different color. Human cytogeneticists adopted chromosome painting as a valuable technique for identifying and characterizing translocations, as well as instances of extra or missing chromosomes.[61]

Cytogeneticists also applied FISH to the analysis of human cells during interphase. At this point in the cell cycle, cytogeneticists believed that the resolution of chromosomal analysis would be greatly enhanced, because the continuous DNA strands making up chromosomes were much less tightly folded. In effect, this was Jorge Yunis's "high-resolution" cytogenetics, described in chapter 3, taken to the next level. Cytogeneticists demonstrated that the ordering of probes could be more precisely distinguished during interphase because chromosomes were much more dispersed. They used a different-colored fluorescent tracer for each probe, which aided in determining relative localization. The development of "interphase cytogenetics" in the late 1980s further advanced the resolving power of human gene mapping and brought researchers slightly closer to bridging the gap between chromosomes and genes.[62]

Over the following years, medical geneticists used interphase cytogenetics to develop simpler approaches for diagnosing microdeletion disorders such as Prader-Willi and Angelman syndromes, which did not require the direct visualization of chromosomes. With interphase cytogenetics, any cell could be examined by human cytogeneticists for certain mutations, instead of the select few that were in metaphase at the time of analysis. The next two chapters explain how FISH and other approaches developed in the 1990s as part of the new hybrid field of "molecular cytogenetics" became valuable in the time-sensitive context of prenatal diagnosis.[63] Molecular cytogeneticists' work, in combination with the growing chromosomal infrastructure for diagnosis, greatly extended the scope of prenatal prevention in the decades ahead.

Conclusions

Michel Morange, in his 1998 book *A History of Molecular Biology,* commented that during the 1980s cell biology developed a "successful cohabitation with molecular biology," rather than facing an existential threat from the more reductive "molecular vision" of the era. Morange pointed to the development of immunofluorescent techniques as central to this synergy, highlighting their use in the study of cell structure and intracellular "traffic."[64] This chapter has examined the ways in which immunofluorescence and other related techniques contributed to the expansion of human cytogenetics during the era of recombinant DNA. In line with Morange's argument that "the molecular vision does not replace previous visions but puts them in a new light," molecular tools became a valuable addition to the approaches and interests of cytogeneticists, rather than a threat to the livelihood of the field.

During the 1980s, as the clinical aims of cytogeneticists and molecular biologists became increasingly intertwined, the field of human cytogenetics was molecularized.[65] While RFLP markers offered new molecular-level footholds, cytogenetic techniques continued to play a central role in human gene mapping, because chromosomes remained the primary organizational units of medical geneticists. Molecular approaches did not eclipse cytogenetic understandings of the genome but instead were integrated into its long-standing chromosomal ways of seeing. The association of Huntington's disease with a RFLP marker was a significant accomplishment, but alone it fell short of medical geneticists' larger goal of mapping Huntington's disease to a chromosomal address. The same was true of other molecular disease findings during this era: they became meaningful in medical genetics not because of their linkage to a molecular marker alone, but following their localization within the genome's morbid anatomy and chromosomal infrastructure for diagnosis. Indeed, as historian Horace Judson noted in 1992, "Today, whenever a gene gets cloned the geneticist has not completed its characterization until he determines which chromosome is carrying it and where."[66]

While molecular biology played an increasingly prominent role in medical genetics and human gene mapping during the 1980s, throughout the decade and into the next, there was an ongoing synergy between chromosomal analysis under the microscope and the molecular approaches of "the new genetics." Human cytogeneticists faced clear limitations in resolution; however, molecular analysis of the human genome in the absence of chromosomal referents also had limited value. The emergence of the hybrid area of molecular cytogenetics reflected the necessity of collaboration between these medical genetics subspecialties, which by the 1980s could no longer afford to occupy distinct domains. Chapter 5 examines the application of molecular cytogenetics in medical genetics during the 1990s and traces a specific instance in which the results of this analysis complemented clinical observations and played a central role as the life histories of two previously distinct disorders became one.

Chapter 5

Institutionalized Disorders

The life histories of genetic disorders have long been shaped by medical expectations about who would be affected and how. As new disorders were delineated and studied during the postwar period, physicians, through their training and textbooks, came to regard certain conditions as being lethal during pregnancy or in early childhood. Given these expectations, physicians were unlikely to diagnose Edward syndrome (trisomy 18) in a teenager, since the disorder had been established as one that caused death before birth or within the first few months of life. In some instances, however, the assumed lethality of a disorder was altered by new treatments, leading to a medical reassessment of the condition's natural and clinical histories. This chapter describes changing medical perceptions of the life history of a syndrome that physicians initially understood as involving deadly birth defects and little else. When new clinical approaches prevented infant lethality, the disorder was recharacterized, and its life history came to be defined by an alternative and much broader pattern of malformations.

Understandings of syndromes evolved in medical genetics based on the observation of developmental patterns in the clinic and genetic markers in the laboratory. Postwar physicians understood syndromes to have variable clinical expression. They were thus diagnosed by referring to a list of potential symptoms, not all of which were expected to occur in any one patient. In addition, physicians recognized that syndromes were likely to show variable clinical manifestations not just among different patients, but also over the life span of affected individuals. This variability complicated clinical diagnosis and categorization. Responding to these perceived challenges, medical geneticists pointed to the identification of genetic mutations as a means for improving decisions about which patients should be "lumped" together under one name, and which should be "split" into other categories.[1]

Victor McKusick began writing about the challenges of syndrome delineation in the late 1960s; he argued that mutations were more objective markers than clinical features for delineating syndromes. An early proponent of the one mutation–one disorder ideal, McKusick claimed, "Medical geneticists have been, and continue to be, lumpers to the extent that they pull together the pleiotropic [variable] manifestations of genetic syndromes." This, he suggested, had not always been the case in past decades, when "medical genetics suffered from excessive and improper splitting."[2] McKusick blamed these instances of inaccurate splitting on the limited focus of clinical specialists, who highlighted different aspects of patients affected by the same disorder. Without genetic mutations to look to as arbiters of syndrome categories, clinicians relied primarily on their specialized interests to define new, seemingly distinct disorders.

The focus of the current chapter is the life histories of two initially distinct disorders that, during the closing decades of the twentieth century, came to be understood as one and the same. The clinicians who first described DiGeorge and velo-cardio-facial (VCF) syndromes identified these disorders from the perspectives of differing places, times, and specialties. When originally delineated, DiGeorge and VCF syndromes appeared to be quite distinct. School-aged children with a relatively minor pattern of clinical malformations made up the initial population of VCF patients, whereas DiGeorge syndrome patients died in the first weeks or months of life as a result of severe immune system and heart defects. Because of the vastly different lethality of these disorders, physicians did not initially interpret their life histories as overlapping.

During the 1970s and 1980s, physicians greatly altered the life history of DiGeorge syndrome through the development of new surgical procedures that corrected the immune system and cardiac defects that had made the disorder fatal during infancy. As the potential life span of DiGeorge patients was extended from months to decades, a new developmental pattern and set of clinical concerns emerged. Something similar occurred in the case of other disorders, such as Marfan syndrome, as described by Reed Pyeritz, who was a medical geneticist and a student of McKusick's: new treatments began to facilitate the examination of natural histories that had never previously been observed.[3] More years of life also brought a greater likelihood of clinical variability. As new treatments extended the life span of DiGeorge patients, clinicians saw a new pattern of malformation, resembling disorders that were not previously diagnosed until later in childhood, adolescence, or adulthood.

Working with the one mutation–one disorder ideal, medical geneticists saw the identification of a common mutation among affected patients as the most compelling evidence that two clinically distinct disorders were in fact one. A common deletion at the chromosomal address 22q11 became a central component of both DiGeorge and VCF syndromes in the 1990s, to the extent that the two disorders came to be accepted as one by physicians, geneticists, and patients. However, despite efforts by various interest groups, coupled with the compelling power of a common genetic mutation, the medical, social, and institutional identities of these historically distinct disorders did not so easily collapse into one. As this representative case shows, the life histories of postwar genetic disorders were much larger and more complex than their etiologies.

In the midst of the Human Genome Project, anthropologist Paul Rabinow coined the term *biosociality* to describe the potential for patients and institutions to come together based on a common genetic mutation. Subsequent scholarly examinations of how social groups develop around disorders, including the DiGeorge and VCF syndromes, revealed a more complicated story.[4] In the course of analyzing the evolving life histories of DiGeorge and VCF syndromes, in terms of both clinical and chromosomal markers, this chapter examines the roles of various individuals and institutions in shaping ongoing discussions about the naming and spectrum of these disorders as both syndromes became associated with the same cytogenetic mutation. The term *DiGeorge syndrome* came under attack early in the disorder's life history and largely fell out of favor after the 1990s for various reasons. Making sense of this trend, I argue that the hierarchy of disorder naming and status in postwar medical genetics was dismissive of etiological heterogeneity and instead was built around the one mutation–one disorder ideal, embodied by the genome's morbid anatomy and chromosomal infrastructure for diagnosis.

The next section looks at the clinical delineation of DiGeorge syndrome and subsequent disagreements over whether the first 100 diagnosed cases represented a distinct syndrome or a pattern of malformation with many genetic causes. In this connection, I trace efforts in the early 1980s to identify a chromosomal mutation for DiGeorge syndrome that would facilitate its isolation in "pure culture," following the model of the 15q11–13 deletion in Prader-Willi syndrome.[5] Next, the independent clinical delineation of VCF syndrome is considered, along with its eventual association with the DiGeorge syndrome deletion and debates among medical geneticists over how to name disorders associated with mutations at the chromosomal address 22q11. Throughout the chapter, I highlight the impact of the one mutation–one disorder ideal on the establishment of distinct genetic disorders, as well as the limitations of this model's influence within the larger social and institutional landscape of medical genetics.

Explaining Early Infant Death

In 1965 Angelo DiGeorge, a pediatric endocrinologist at St. Christopher's Hospital for Children in Philadelphia, commented in the *Journal of Pediatrics* on three infants he had cared for, over a period of six months, who showed notably similar developmental patterns. Each of the infants came to medical attention because of hypoparathyroidism (reduced activity of the parathyroid glands), which caused low blood calcium levels. All three died in the first year of life. DiGeorge noted that several similar patterns of malformation involving hypoparathyroidism, recurrent infections, and early infant death had been described over the previous century. Patients in some of the earlier cases also lacked the thymus, an organ located above the heart that develops from the same fetal structures as the parathyroid glands. Absence of the thymus could not be confirmed before autopsy, but in each of the three cases DiGeorge had observed, he was excited to learn that thymic tissue had not been found during postmortem examination.[6]

A fourth infant with a similar developmental pattern involving hypoparathyroidism and recurrent infections was admitted to St. Christopher's Hospital in January 1965. DiGeorge saw that this patient offered an exciting opportunity to actively examine thymus function in humans, which remained poorly understood. He later recounted, "Here was the perfect experiment of Nature for the thymologist." Interestingly for DiGeorge, blood work on each of these four similarly affected patients had found normal levels of antibodies, suggesting to him that existing hypotheses about the thymus's role in the immune system were incorrect. Over the next 10 months, DiGeorge performed tests and experiments on the fourth patient, including lymph node biopsies, sensitivity tests to various antigens, and the placement of a skin graft from an unrelated adult. Each demonstrated that his immune system was not responding to foreign substances, despite possessing normal levels of antibodies. The boy suffered from recurring infections. He never left the hospital and died after 16 months. An autopsy found no thymus.[7]

In a follow-up report, DiGeorge noted that the hypothetical possibility of treating this fourth infant with a thymus transplant had been debated but never attempted. He pointed to the lack of available embryonic thymus tissue for transplant, as well as continued uncertainty about whether the patient's condition actually resulted from the absence of a thymus. The clinicians involved also recognized that such a treatment would compromise their experimental findings. As DiGeorge put it, "If our patient did indeed have no thymus it was important not to confuse the findings by therapeutic experiments."[8] The St. Christopher's physicians did not expect a child with this condition to survive for more than a few months, but they were interested in learning as much as they could before he died. In these years before the establishment of institutional bioethics oversight, which more explicitly divided the roles of patient and experimental subject, individual physicians were left to manage the balance between clinical care and research aims.[9]

Based on his observational and experimental work, DiGeorge inferred that the developmental pattern he had observed in these four patients likely represented a unique clinical disorder. Acknowledging his role in bringing this condition to wider medical attention during the late 1960s, many physicians, following the suggestion of immunologist Robert Good, began referring to congenital lack of the thymus, and its clinical implications, as "DiGeorge syndrome." The adoption of eponyms to name disorders had been an important part of the awards system of medicine since the nineteenth century and was particularly popular in the decades after World War II.[10] While "flattered" by the suggested eponymous designation, DiGeorge "demurred," pointing out that such cases had been reported as early as the first half of the nineteenth century. He quoted an 1829 letter by physician Henry Harington, published in the *London Medical Gazette,* describing congenital absence of the parathyroid glands and thymus.[11] Protestations aside, the term *DiGeorge syndrome* was widely adopted by clinicians. The significance of DiGeorge's contribution was further strengthened by reports of additional similarly affected individuals in the medical literature over the next years. Among these patients, physicians reported a range of symptoms, including seizures, palate abnormalities, developmental delay, and heart defects.[12]

Moving beyond the theoretical implications of this disorder, some clinicians also worked to develop treatments for DiGeorge syndrome using thymus tissue transplants to restore immune function. In 1968 a group of pediatric researchers at the University of Miami School of Medicine reported that the transplant of fetal thymic tissue to an infant had successfully restored immune function. More than a year later, the infant remained free of infections. Another report from Harvard University simultaneously described the successful transplantation of fetal thymic tissue in an infant with DiGeorge syndrome, who, after 18 months, also showed reconstituted immune function. Just as DiGeorge had described, these infants possessed a normal level of antibodies in their blood. Researchers believed that the thymus transplant reestablished an important second step in the immune response, facilitating the recognition of novel antigens. Without a thymus, they argued, the body could not produce antibodies targeting new infections.[13]

An Expanding Clinical Designation

Thymus tissue transplants offered hope for the longer-term survival of some DiGeorge syndrome patients. However, as medical familiarity with this pattern of malformation led to the diagnosis of more cases in the 1970s, physicians also reported that heart defects were a common cause of infant death in this population. The vast majority of the first 54 patients diagnosed with DiGeorge syndrome also showed some heart defect. While physicians lumped all of these patients under the category of DiGeorge syndrome, their symptoms were variable in expression. Many individuals had a small thymus rather than complete absence, a condition that physicians called "partial DiGeorge syndrome." Some of the less severely affected patients were living into their teenage years and beyond, a very different circumstance from what DiGeorge had described.[14]

The clinicians who examined these patients tended to see, study, and write about the malformations that they found most interesting. While DiGeorge was drawn to the implications of an absent thymus and the resulting immune deficiency, physicians in other specialties focused on different defects. As DiGeorge readily acknowledged, a decade before his report, New York University physician David Lobdell had described a patient with similar characteristics, but Lobdell had focused on the absence of parathyroid glands.[15] In 1977 another team of pediatricians, in the cardiac division at Montreal Children's Hospital, reported on seven patients, all but one of whom died in the first weeks of life from cardiac problems. These clinicians claimed that "diagnosis of DiGeorge syndrome should be possible in the newborn. The important features are not, however, related to immune deficiency, but rather to severe congenital cardiovascular disease."[16]

What, then, were the most common and deadly features of this disorder? In an attempt to circumvent the biased interests and patient populations of different specialists, pediatricians at the University of Washington performed a retroactive study of patient records to identify potential cases of DiGeorge syndrome. They included only patients who had at least two of the three conditions normally associated with the disorder: partially or completely absent thymus, heart defects, and hypoparathyroidism. In total, 24 patients were identified among 3,469 autopsy records. One living patient, age seven, was also included. Fifteen of these individuals first came to medical attention immediately after birth owing to congenital heart disease. The same number died in the first month of life, and nine more died before age one. Low blood calcium and seizures were common, as were recurrent infections, but these symptoms became an issue only in patients who survived longer than a week. Even in a retroactive population, the malformations that most directly affected survival in DiGeorge syndrome seemed to vary among patients.[17]

In the 1970s a diagnosed case of DiGeorge syndrome involving two or more major malformations led to early infant death in the vast majority of instances. However, as the University of Washington team reported following their 1979 retrospective study, patients diagnosed with what some called partial DiGeorge syndrome often lived much longer. These instances provided opportunities to follow the disorder's long-term "natural" history. Many partial DiGeorge syndrome patients suffered from low blood calcium in infancy, but it sometimes resolved in childhood. Another feature that physicians noted as less severely affected individuals grew older was the incidence of mild to moderate mental retardation. These findings raised new concerns about the well-being and care of DiGeorge patients.[18]

The expanding DiGeorge "spectrum," as University of Utah dysmorphologist John C. Carey characterized it, led some physicians to begin questioning whether all of the patients that had been diagnosed under the label of DiGeorge syndrome were affected by the same disorder. In 1980

Carey wrote a letter to the editor of the *Journal of Pediatrics*. He responded to the University of Washington report by suggesting that what had been treated as a single entity under the designation DiGeorge syndrome might in fact represent a "complex" of related developmental defects that could be "one feature of several distinct syndromes."[19] Carey proposed that there might exist a distinct DiGeorge syndrome among this population, but that most of the diagnosed individuals probably did not have it. The more distinctive syndrome would include the DiGeorge "complex" plus additional characteristics. Carey put forth "the abnormal facial features noted in some children" as potentially demarcating what could be "labeled the DiGeorge *syndrome*."[20]

During the 15 years after DiGeorge's initial report, physicians diagnosed about 100 individuals with DiGeorge syndrome. Over this period, however, what counted as DiGeorge syndrome was increasingly a matter of debate among medical geneticists. Syndromes were understood to involve a constellation of symptoms that need not all be present in any one affected individual. As pediatrician Kurt Hirschhorn put it, identifying unique syndromes was often "tricky business" because "the conception of a syndrome is simply that, once it's described, you have the possibility of recognizing it and accepting the fact that the same syndrome can have manifestations that are absent or present."[21] The syndrome concept allowed for a lot of tolerance in terms of clinical variability: even the most definitive features need not be seen in all patients. This understanding of syndromes fit quite comfortably with the one mutation–one disorder ideal of medical genetics, because it allowed for the variability in clinical outcomes that physicians often observed in patients with the same mutation.

Medical geneticists, however, were concerned about the erroneous "lumping" of conditions that had similar clinical features but heterogeneous genetic causes.[22] In his letter to the editor, Carey opposed using the term *syndrome* to describe patterns of malformation that had been identified in multiple disorders. Medical geneticists like Carey, who had trained in the subspecialty of dysmorphology with David Smith, understood syndromes as distinct entities with just one cause, often genetic.[23] Most of the cases of DiGeorge syndrome that physicians had diagnosed since 1965, Carey argued, were likely part of a larger, genetically heterogeneous group of malformations, and not a single syndrome. From a medical genetics perspective, then, resolving the true nature of DiGeorge "syndrome" required identifying its genetic cause(s). For this, medical geneticists during the 1970s and early 1980s turned to chromosomal analysis.

Isolating Syndromes in "Pure Culture"

From its origins in the 1930s, the field of medical genetics was attracted to developing a "germ theory of genes." As historian Nathaniel Comfort has insightfully described, the means of searching for the "germs" that caused genetic disease shifted to the analysis of chromosomes with the rise of human cytogenetics in the 1960s.[24] As Victor McKusick expressed it, chromosomal analysis offered medical geneticists the opportunity to isolate and study conditions in "pure culture" within the genome's morbid anatomy.[25] In much the same way that nineteenth-century microbiologists used Koch's postulates to culture the bacterial causes of disease and study these "germs" in isolation, chromosomal analysis offered the chance to examine the variable clinical expression of a disorder in individuals who were isolated out based on a shared genetic mutation.

Efforts to identify a common chromosomal abnormality among clinically variable DiGeorge syndrome patients were as old as the disorder, beginning in Philadelphia at St. Christopher's Hospital. Philadelphia was a major center of human cytogenetics research at this time, and at St. Christopher's, PhD cytogeneticist Hope Punnett was part of the team examining DiGeorge syndrome. However, no cytogenetic abnormalities were identified in this case.[26] Beyond this, chromosomal analysis was rarely reported in DiGeorge syndrome patients during the 1960s and 1970s. The reasons for the lack of cytogenetic examination are unclear, but it may have resulted from the lack of expertise at many locations where the disorder was diagnosed.

Even in the absence of widespread cytogenetic examination at this time, clues about the genetic origin of DiGeorge syndrome did begin to emerge. In 1972 pediatricians at Georgetown University reported on a family that was affected by thymic abnormalities. The mother, who possessed a partial thymus and deficient parathyroid glands, had given birth to two children over five years who lacked the thymus and parathyroid glands. Both children, a boy and a girl, had died in early infancy. These physicians believed that each of the family members had been affected by a common pattern of malformation, which they called DiGeorge syndrome, and suggested that the disorder had been inherited as a dominant Mendelian trait.[27]

A second report of a family that appeared to be affected by DiGeorge syndrome came in 1981 from the Children's Hospital at the University of Helsinki in Finland. In this family, both parents were reported to be normal, while three of their four children, all girls, were diagnosed with variable forms of DiGeorge syndrome. All three children had a common heart defect and little or no thymic tissue. The eldest girls died within months owing to cardiac complications, but an operation saved the youngest child, who had a small but functional thymus and no obvious immune deficiency. Because three children in this family had DiGeorge syndrome, but their parents were clinically normal, the physicians suggested a recessive inheritance pattern.[28]

A second team of Finnish clinicians was examining yet another case of apparent DiGeorge syndrome at this time. The affected family came to the attention of physicians in the northern city of Oulu in 1980, when a newborn daughter was found to have a severe cardiac defect. The girl died three weeks after birth, and an autopsy revealed that she had lacked a thymus and parathyroids. Physicians learned from this family that two siblings and a cousin of this girl had also died in the first months of life. Each had shown a similar pattern of malformation. Additional infant deaths were reported in the children's and parent's generations, but no details were available. Once again, the finding that DiGeorge syndrome had affected multiple individuals in this extended family led physicians to infer an inherited genetic cause.

Clinicians in Oulu sent tissue samples from this family to be studied by human cytogeneticists at University of Helsinki. The laboratory, which was led by medical geneticist Albert de la Chapelle, was actively engaged in mapping human disease genes using cytogenetic methods. Examining the cells of this infant girl, de la Chapelle's group identified an unbalanced translocation involving chromosomes 20 and 22. Some genetic material was deleted on 22, and some was duplicated on 20. Luckily, cytogenetic analysis had previously been performed on other affected family members, and microscopic slides were still available for reexamination. A similar translocation was identified in each infant who had died of apparent DiGeorge syndrome, and the translocation was traced back to siblings who were the parents of the affected children. The parents had a balanced form

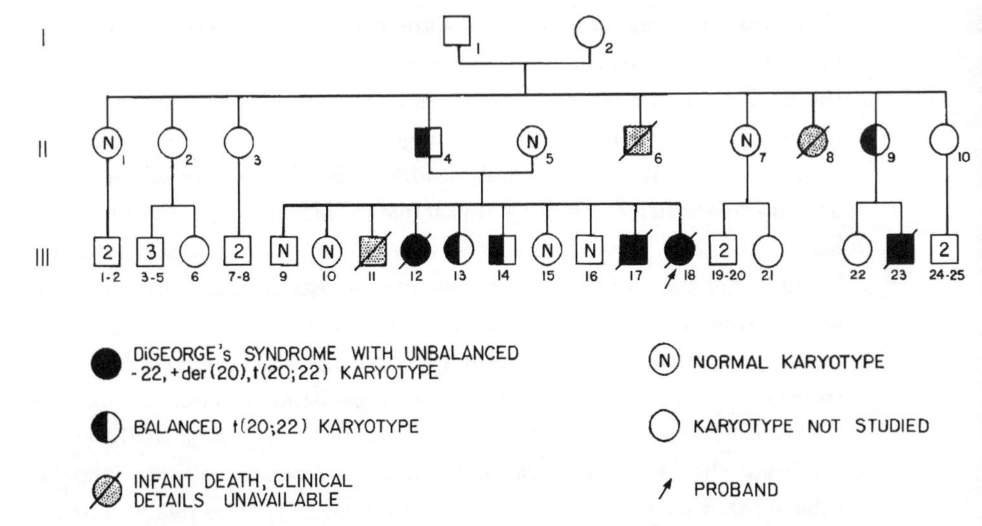

Pedigree showing the inheritance of DiGeorge syndrome in an extended family, resulting from an unbalanced translocation involving the deletion of genetic material from chromosome 22. *Source:* Figure 1 in A. de la Chapelle et al., "A Deletion in Chromosome 22 Can Cause Digeorge Syndrome," *Human Genetics* 57, no. 3 (1981): 254. Reprinted with permission from Springer Science+Business Media. Copyright Springer-Verlag 1981.

of this translocation, with no added or missing genetic material, and were clinically normal. De la Chapelle concluded from these findings that either a deletion on chromosome 22 or duplication of chromosome 20 caused DiGeorge syndrome. "The implication," he wrote, "is that in this family, the disease is caused by the chromosomal aberration."[29]

De la Chapelle's laboratory also performed chromosomal analysis on the other DiGeorge syndrome family that had been recently identified in Helsinki. Karyotypes from the three affected children and both parents showed no abnormalities. For additional insights on the implications of the unbalanced translocation, de la Chapelle instead looked to previous reports of the clinical impacts associated with trisomy 20 and monosomy 22. Deleted material from chromosome 22 had been linked in earlier studies with features similar to DiGeorge syndrome. From this, de la Chapelle inferred that the deleted material in their own patient, which spanned from the tip of the short arm (22p) through the long arm location 22q11, was probably the site of the causative DiGeorge "gene deletion."[30]

At this time, medical geneticists knew of multiple ways that chromosomal translocations could cause disorders: by disrupting a single gene at the breakpoint, by recombining genes from multiple chromosomes, and through the deletion or duplication of genetic material. Both clinical and cytogenetic features of the family studied aided de la Chapelle's group in determining which mechanism was active in this DiGeorge syndrome. He later explained, "The reason why we were able to guess that this was a deletion rather than a translocation breakpoint that affected a [single] gene was two-fold. First, [there was] very great variability in clinical features of DiGeorge syndrome. . . . The second thing was that there were balanced translocation carriers, like the father of the proband, who were normal. So the translocation breakpoints did not affect the gene."[31] Because DiGeorge syndrome showed so much variability in clinical outcomes, de la Chapelle inferred that more than one gene on chromosome 22 was likely to be involved in causing it. In addition, his family study had revealed that both normal and affected individuals had the same chromosomal abnormality. The disorder appeared only when this translocation caused the loss of genetic material. A decade later, this cytogenetic explanation was accepted as correct.

De la Chapelle's cytogenetic studies of DiGeorge syndrome families during the early 1980s followed a model provided by Vincent Riccardi, who in 1978 reported on a common deletion on chromosomal 11 in aniridia–Wilm's tumor patients. Riccardi's laboratory drew on previous reports of a translocation involving this chromosome in affected patients and had applied high-resolution analysis in the effort to identify a causative mutation. De la Chappelle's work on DiGeorge syndrome occurred at the same time as Riccardi and David Ledbetter's analysis of chromosome 15 in Prader-Willi syndrome, described in chapter 3, which also followed the identification of a translocation in Prader-Willi patients. Each group was at the forefront of using banded chromosomal analysis to isolate genetic syndromes in "pure culture" so as to improve their clinical characterization and delineation.[32]

While Ledbetter succeeded, using high-resolution techniques, in identifying a Prader-Willi deletion, de la Chapelle's team had no such luck in identifying a discrete chromosomal mutation for DiGeorge syndrome. Nonetheless, their work suggested a possible genomic location for the disorder on chromosome 22, and it laid the foundation for more targeted work on other patients. Though de la Chapelle's group was actively involved in disease-gene mapping, they chose not to pursue DiGeorge syndrome any further after 1981. The laboratory's research primarily focused on genetic disorders that were native to Finland, and, like many of their peer groups at this time, they were in the early stages of transitioning from cytogenetic to molecular approaches. Over the next decade, other research groups used these molecular tools to identify a discrete chromosomal address for DiGeorge syndrome, seeking its isolation in "pure culture."[33]

Locating DiGeorge Syndrome

De la Chapelle's report of a chromosomal abnormality in families affected by DiGeorge syndrome immediately drew the attention of medical geneticists in Philadelphia. Although DiGeorge did not remain actively involved in research on the syndrome that had been named in his honor, other clinicians in Philadelphia were aware of the disorder and had identified it among their own patients. Earlier in the 1970s, medical geneticist Beverly Emanuel had developed chromosomal banding techniques for clinical application at the University of Pennsylvania and the affiliated Children's Hospital of Philadelphia. She eventually began to specialize in abnormalities of chromosome 22 and had identified the cause of another rare disorder involving the duplication of a small portion of the chromosome, which was later named after her.[34]

Around 1980, a child with DiGeorge syndrome was clinically diagnosed at the Children's Hospital of Philadelphia. Emanuel's cytogenetic analysis of the individual identified a missing portion of chromosome 22, owing to a translocation with chromosome 10. She submitted the finding for presentation at the American Society of Human Genetics shortly before de la Chapelle's report was published. As Emanuel put it, "It was almost like an 'ah ha' moment. We have a child here missing part of 22 who has DiGeorge, and they [de la Chapelle's group] have a family with three affecteds missing part of 22 with DiGeorge."[35] Having identified this similarity, Emanuel reached out to DiGeorge and Punnett across town at St. Christopher's Hospital to ask whether they had found instances of DiGeorge syndrome patients with chromosomal abnormalities. In fact, they had identified two such patients, both of whom had unbalanced translocations, one involving chromosomes 20 and 22 and the other chromosomes 3 and 22.

Researchers at the Children's Hospital of Philadelphia and St. Christopher's collaborated in writing up a report of these patients in 1982. All three infants showed heart defects and an underdeveloped thymus. Each had died before age one. Based on these cases and published experience with other deletions involving chromosome 22, the researchers suggested that the critical deleted region among these patients was most likely 22q11. They acknowledged, however, that banded chromosomal analysis in three other similarly affected patients had revealed no translocations or apparent deletions. These inconsistent findings, the Philadelphia group suggested, highlighted the need for high-resolution chromosomal analysis in the study of DiGeorge patients, as had been successfully applied by Riccardi and Ledbetter to make visible deletions associated with retinoblastoma, aniridia-Wilms tumor, and Prader-Willi syndrome.[36]

An addendum to the collaborative report out of Philadelphia referred to another family that had very recently been identified by physician Frank Greenberg. Greenberg subsequently moved to the Baylor College of Medicine in Houston, where he collaborated with Ledbetter in applying high-resolution cytogenetics to the study of DiGeorge syndrome. An infant in the affected family was described as having mild heart defects, abnormal facial features, and immune deficiency. With treatment, the child developed normally. A younger sibling of his had died at two months of age and was reported to have had a heart defect as well as an underdeveloped thymus and underdeveloped parathyroids. Examination of their mother had identified subtle facial abnormalities and mild immune deficiency. Greenberg believed that all three had DiGeorge syndrome.

Ledbetter's cytogenetic studies of the mother and her living child revealed an unbalanced translocation involving chromosomes 4 and 22, leading to the potential deletion of genetic material in the 22q11 region. Based on these findings, Ledbetter and Greenberg wrote in 1984, "We must consider the possibility that some cases of DiGeorge syndrome might have an interstitial deletion of 22q11, a situation analogous to Prader-Willi syndrome."[37] Influenced by the representative case of a Prader-Willi deletion, Ledbetter and Greenberg examined additional DiGeorge patients, looking for deletions within otherwise normal-looking copies of chromosome 22. However, they were not successful in visualizing an analogous DiGeorge syndrome deletion.

The inheritance of DiGeorge syndrome in multiple families, along with the recurrent inability to make visible an associated chromosomal deletion,

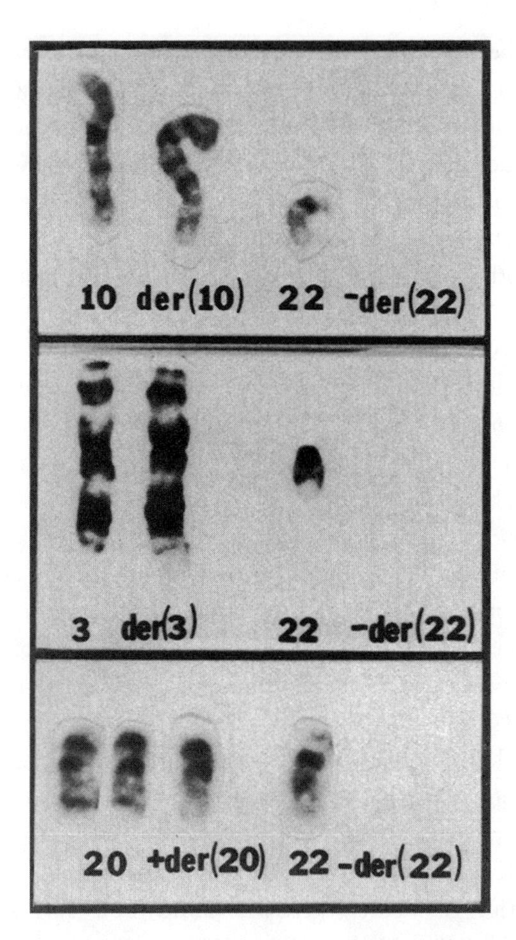

Multiple instances of unbalanced translocations involving chromosome 22 from patients with DiGeorge syndrome. In each case, some genetic material from chromosome 22 was deleted. *Source:* R. I. Kelley et al., "The Association of the Digeorge Anomalad with Partial Monosomy of Chromosome 22+," *Journal of Pediatrics* 101, no. 2 (1982): 198. Copyright 1982. Reprinted with permission from Elsevier.

led human cytogeneticists in the mid-1980s to increasingly think that the disorder might be caused by a molecular mutation in one gene, rather than a chromosomal aberration. Emanuel commented, "At the time, the thinking in the field was, somewhere on 22 there's a DiGeorge gene, and probably the patients with the translocation were just putting us into the right region."[38] As in the cases of fragile X and Prader-Willi syndrome, the cytogenetic link between 22q11 and DiGeorge syndrome drew the interest of molecular biologists, leading to significant molecular characterization of this region in the late 1980s and the 1990s.[39] Emerging molecular tools, including restriction enzyme analysis and FISH, contributed over subsequent years to the identification of a common mutation in DiGeorge syndrome patients. Notably, the mutation was also found among patients who were diagnosed with another disorder that had its own distinct life history.

A Syndrome in School-Aged Children

In 1974 Robert Shprintzen, a PhD speech pathologist, was hired to direct the Center for Craniofacial Disorders at the Montefiore Medical Center in New York City. The center served about 500 patients annually from the Bronx and, based on referrals, from throughout the region. Soon after his arrival, Shprintzen received a phone call from nearby Sarah Lawrence College asking if he would be interested in training students from its new, first-in-the-nation, genetic counseling program. Shprintzen did not know exactly what genetic counseling was, or how it might contribute to his center, but he accepted the offer for additional, and free, expertise. The center soon welcomed Rosalie Goldberg, who went on to work there for 25 years.[40]

With the addition of genetics expertise, Shprintzen began to engage with the developing field of dysmorphology. He invited major figures from the subspecialty, such as David Smith and Robert Gorlin, to spend time in the clinic and train its fellows, including Shprintzen himself, who had no formal education in medical genetics. Smith and Gorlin, along with Gorlin's research fellow Michael Cohen, became important mentors and colleagues of Shprintzen. Over the years, through phone conferences and annual visits to the center, they aided Shprintzen in delineating new genetic disorders and proved to be valuable allies as he built his professional identity.[41]

During Shprintzen's early years at Montefiore, the Craniofacial Center received a continuous flow of referrals involving children with various head, face, palate, and speech abnormalities. One of these patients was a young girl who was referred because of her hypernasal speech, resulting from malformation of the palate. She was also affected by learning disabilities and had subtle facial features that would have been overlooked by most but were notable to Shprintzen, after his informal training in dysmorphology. As he examined similar patients over the coming months, Shprintzen observed among a small group of them common facial features that to him made these children look more like one another than like their individual parents. Shprintzen described each as having "a large, fleshy nose with a broad nasal ridge," flattened cheeks, and narrow eye openings. Nine of the 12 patients also had heart defects, and 11 showed learning disabilities. While concerning for families, this condition was relatively mild. Each of the children lived normally at home and, in all cases except one, attended school in regular classrooms with extra assistance.[42]

After his training in malformation patterns from Smith and Gorlin, Shprintzen was acutely aware of the significance of finding a patient cohort with similar features. He explained,

> I probably had about a half dozen or so patients who walked through the door within a few months of each other, who had remarkably similar presentations. . . . We were so sensitive to the issue of multiple anomaly disorders because David Smith had always said that if you are seeing something that looks like the pattern, you have to be able to recognize the pattern and think in terms of not individual features but rather the pattern of features you see, with some variability. So after the sixth or so I said, "This is more than just a coincidence," and that's when we got to work writing up our paper that was published in 1978.[43]

While their parents surely saw in the faces of these children features resembling other members of their nuclear family, Shprintzen, as a dysmorphologist, was more stuck by characteristics that allowed him to categorize the children as part of a clinical family, all having the same disorder.

In line with Smith's advice, Shprintzen's report described "a pattern of similarities among 12 patients which are felt to represent a *newly recognized congenital malformation syndrome*." He coined the descriptive name velo-cardio-facial (VCF) syndrome for the disorder, because its most notable features were inborn oral and nasal (velopharyngeal) abnormalities, heart defects, and distinct facial characteristics. Shprintzen was careful to point out that he believed this to be a distinct and newly defined clinical syndrome. He argued that the unique combination of abnormal cardiac and facial features, along with learning disabilities, made the disorder unlike any other that had previously been clinically delineated.[44]

Dysmorphologists held that some patterns of malformation resulted from a single physical defect, while others involved systematic effects throughout the body. As Shprintzen saw it, the spectrum of symptoms he had reported indicated more than a localized developmental defect. In 1978 he noted, "The rather unusual combination of symptoms in the twelve patients described above becomes difficult to explain in relation to the timing of fetal development." The disorder, he argued, should be considered a multisystem syndrome, rather than an isolated fetal defect. Shprintzen did not identify a specific genetic or environmental cause for this pattern of malformation; however, he did note one instance of apparent familial transmission.[45]

By 1981 Shprintzen had expanded his VCF syndrome patient population to 39. Most had been referred to and diagnosed at the Montefiore Craniofacial Center because of their hypernasal speech. Nine of these cases were identified in other locations based on Shprintzen's 1978 report. Four instances of familial transmission of VCF syndrome were noted, suggesting that the syndrome might be inherited in a dominant manner. Chromosomal analysis found no visible abnormalities. The next year, Smith included "Shprintzen syndrome" in *Recognizable Patterns of Human Malformation* and identified it as a "probable autosomal dominant." VCF syndrome became an increasingly well-established disorder in clinical texts during the 1980s. However, it was regarded as a difficult diagnosis, owing to significant clinical overlap with other disorders such as fetal alcohol and DiGeorge syndromes and the lack of a cytogenetic marker.[46]

The identification of a growing population of VCF patients at this time also led to the recognition of new symptoms and of increasing clinical variability among those diagnosed. In a 1985 conference presentation, Shprintzen's medical fellow Robert Marion spoke of a significant phenotypic overlap between what he termed "the DiGeorge sequence" and VCF syndrome. After one patient diagnosed with VCF syndrome was found to have a type of immune dysfunction common in DiGeorge patients, researchers had examined 60 other VCF syndrome patients and found that many had an underdeveloped thymus. The parents of these children also reported reoccurring infections, and some others were found to have DiGeorge-like immune dysfunction. Moving in the other direction, it was noted that various clinically diagnosed DiGeorge patients had facial features that were common in VCF syndrome. Since DiGeorge and VCF syndromes both occasionally showed autosomal dominant inheritance patterns, Marion suggested that "VCF should be considered in any familial instance of DGS." DiGeorge "sequence," he argued, not only resembled VCF syndrome but in some cases might be a component of VCF syndrome.[47]

At this time the accuracy of the term *DiGeorge syndrome* was being called into question in multiple circles. In 1986 medical geneticists Edward J. Lammer and John Opitz published an article in the *American Journal of Medical Genetics,* arguing that DiGeorge syndrome would be more accurately classified as a "developmental field defect." The condition was presented not as a stand-alone syndrome but as a pattern of malformation resulting from a single developmental anomaly, which "occurs as a recognizable dysmorphogenetic unit in several causally different malformation syndromes."[48] In effect, the authors suggested that this malformation pattern was not a syndrome, because it appeared to have multiple causes and was the result of other disorders rather than a unique genetic entity. Opitz, the founding editor of the *American Journal of Medical Genetics* and a well-established guru of disease classification, had previously been critical of the "indiscriminate use" of the term *syndrome* "by sociologists and political commentators." He argued that the designation *syndrome* should be used only to describe ontologically distinct disorders, in the Linnaean tradition of distinguishing species.[49]

John Carey, who in 1980 had called into question the designation DiGeorge syndrome, reported a decade later on a family that came to him for genetic counseling after losing an infant to the disorder. Carey noted that the child's father had been born with a cleft palate, continued to have hypernasal speech, was affected by learning disabilities, and had a facial appearance similar to that of VCF syndrome. It appeared to Carey that this man had the disorder and had passed it on to his infant son, who was initially diagnosed with DiGeorge syndrome after dying at three days of age. Carey also presented DiGeorge as something less than a syndrome. He argued, "All previously reported cases of autosomal dominant DiGeorge anomaly are examples of the velocardiofacial syndrome."[50] What had been a lethal disorder among infants was increasingly being seen in the late 1980s as part of the larger life history of another syndrome, which had first been delineated in school-aged children and showed great variability in severity and symptoms.

Notably, Carey had previously left open the possibility that a distinct "DiGeorge *syndrome*" might exist within the broader population of patients that he had designated as having the DiGeorge "complex." In his 1980 report, he had suggested, "If my hypothesis that there is a distinctive face in *some* children is verified, I would suggest that such an individual be labeled the DiGeorge *syndrome*." During the following years, such a disorder was broadly defined, but under the designation VCF syndrome. As it became clear that VCF syndrome was effectively the "DiGeorge *syndrome*" that Carey had pondered, no one stepped in to claim priority for DiGeorge.[51] Included was Angelo DiGeorge himself, who still saw patients with his namesake disorder but had ceased contributing to publications about it in the 1970s. Shprintzen, however, continued to promote and build institutions around his designation, VCF syndrome. The next section describes how the uptake of molecular cytogenetics techniques for studying these disorders in the 1990s offered additional opportunities to solidify a particular designation.

Diagnosis and Prevention

Cytogenetic studies in the early 1980s had suggested an association between DiGeorge syndrome and the chromosomal address 22q11. However, for years no smoking gun, in the form of a deletion at this location, was made visible by cytogeneticists. In 1988 interstitial deletions (which occurred in the middle, rather than at the end, of a chromosome) at 22q11 were reported in 2 DiGeorge patients among more than 30 examined. Using high-resolution analysis, David Ledbetter was able to visualize an interstitial deletion in a patient diagnosed in Houston, and he was called upon to verify a similar deletion first visualized in San Diego. Notably, in the course of these studies, Ledbetter also identified an interstitial deletion at the genomic location 10p13 in a DiGeorge patient. This provided another example of the purported genetic heterogeneity in the cause of this pattern of malformation. Over the following years, additional patients who had been clinically diagnosed with DiGeorge or VCF syndrome were also found to have microdeletions at 10p13, as well as at other genomic locations.[52] Given the variable results from high-resolution cytogenetics and reflecting the evolution of his own interests, described in chapter 3, in the late 1980s, Ledbetter suggested that molecular analysis could improve the detection and characterization of deletions in various syndromes, including Prader-Willi and DiGeorge.[53]

During the late 1980s and early 1990s, Beverly Emanuel was actively developing molecular techniques, described in chapter 4, to examine chromosome 22 for disease-causing mutations. In 1987 her laboratory identified molecular probes that hybridized to the DiGeorge syndrome region of chromosome 22.[54] In subsequent years postdoctoral fellow Deborah Driscoll took up the molecular analysis of DiGeorge syndrome. Driscoll was a physician who had completed her obstetrics residency at the University of Pennsylvania and decided to remain in Philadelphia to do a postdoctoral fellowship at Children's Hospital of Philadelphia, where she spent a year in Emanuel's laboratory.[55] She used the molecular probes in the study of cell lines from 14 patients who had been clinically diagnosed with DiGeorge syndrome. High-resolution cytogenetic analysis had identified a visible interstitial deletion at 22q11 in 5 of these individuals. Driscoll showed in 1992 that two molecular probes that hybridized in the 22q11 chromosomal region in normal individuals did not hybridize in the cell lines from the DiGeorge syndrome patients, suggesting that all 14 had a common deletion.[56]

Aware of the clinical overlap that had been reported between DiGeorge and VCF syndromes, Driscoll was interested in applying the DiGeorge probes to examining chromosome 22 in VCF patients, to see whether she could identify a common deletion. To do this, Driscoll and Emanuel turned to the cleft palate clinic at the Children's Hospital of Philadelphia to inquire about VCF patients. Plastic surgeon Donald LaRossa, who directed the clinic, was somewhat skeptical at this time about VCF syndrome; he felt that Shprintzen was overemphasizing its prevalence. Driscoll and Emanuel were persistent in their interest, however. With the assistance of clinical geneticist Elaine Zackai and genetic counselor Donna McDonald-McGinn, they identified multiple patients who appeared to have undiagnosed VCF syndrome. These children had been largely overlooked by the clinic as cases of VCF and DiGeorge syndromes because they did not possess significant immune system or heart defects. However, they had cleft abnormalities, hypernasal speech, and learning disabilities. As Emanuel put it, "Once we had a way of testing it with a reagent that would work [targeted DNA probes] it was pretty obvious."[57] Individuals with a wide range of symptoms from subtle to life-threatening all appeared to possess a similar deletion on chromosome 22. In line with the one mutation–one disorder ideal of medical genetics, these patients were interpreted as having the same syndrome.

During the early 1990s, Driscoll also worked to develop an approach for the prenatal diagnosis of DiGeorge and VCF syndromes. Because they appeared to be inherited in an autosomal dominant matter, these disorders were strong candidates for targeted prevention in families that already had an affected child. Driscoll advocated for the development of a fluorescence in situ hybridization (FISH) test, suggesting that FISH would "ultimately prove to be the most rapid and efficient method for the antenatal detection of deletions in the diagnosis of DGS [DiGeorge syndrome]." As a model, she pointed to a FISH assay that Ledbetter had developed for the prenatal diagnosis of the microdeletion disorder Miller-Diecker syndrome. Two years later Driscoll reported the successful application of FISH for detecting the 22q11 deletion in children, in adults, and prenatally.[58] DiGeorge and VCF syndrome were thus included within medical geneticists' chromosomal infrastructure for diagnosis and prenatal prevention. While debates continued over what to call the resulting condition, medical geneticists, using molecular cytogenetic approaches, had established the 22q11 deletion as a pathogenic mutation.

Naming Disputes

As their population of clinically diagnosed DiGeorge and VCF syndrome patients expanded in the 1990s, Driscoll and colleagues found that most, but not all, could be shown molecularly to possess a 22q11 deletion. Among 90 patients studied during the early 1990s, 88% of those diagnosed with DiGeorge syndrome and 76% of those thought to have VCF syndrome were found to possess the deletion. Driscoll suggested that the absence of deletions in some patients "may reflect the clinician's ability to diagnose DGS [DiGeorge syndrome] accurately in contrast to their ability to recognize VCFS." If no deletion was detected, it might point to a clinical error in diagnosis, and this appeared more likely to happen for VCF than for DiGeorge syndrome. Driscoll and colleagues said that while a DiGeorge diagnosis "is based on the presence of three findings . . . clinical criteria for establishing a diagnosis of VCFS are not as restricted."[59] What counted as DiGeorge or VCF syndrome ultimately depended on whether clinical or molecular findings were considered to be the most reliable.

Following the molecular work of Driscoll and colleagues, Robert Shprintzen sought to establish VCF syndrome, the clinical designation that he had coined in 1978, as the singular result of a 22q11 deletion. He drew on more than a decade of skepticism about the term *DiGeorge syndrome* to characterize it instead as a "sequence," which sometimes resulted from VCF syndrome. In a 1994 letter published in the *Journal of Medical Genetics,* Shprintzen disagreed with recent reports by Driscoll and others, which had suggested that fewer than 100% of VCF syndrome patients had a 22q11 deletion. He argued that any individual identified as having VCF syndrome and not found to also possess the 22q11 deletion had been misdiagnosed by clinicians owing to insufficient clinical expertise or criteria. He went on to note, "In our experience, clinical application of the diagnosis of velocardiofacial syndrome by careful analysis (preferably longitudinal) of clinical phenotype has led to a 100% accurate detection of a 22q11 microdeletion in all cases."[60] Shprintzen claimed that the clinical experience he brought to VCF diagnosis ensured 100% diagnostic accuracy, which corresponded with 100% presence of the 22q11 deletion. In the process, Shprintzen refuted any suggestion that VCF syndrome was causally heterogeneous. In his rendering, the disorder and the 22q11 deletion were one and the same—an argument rooted in the rhetorical power of the one mutation–one disorder ideal.

Shprintzen also pushed back against the notion that DiGeorge and VCF syndrome, based on the common role of a 22q11 deletion, should be united as one disorder under a new designation. Peter Scambler and colleagues had recently suggested the adoption of the acronym CATCH-22 to capture the various clinical features associated with the 22q11 deletion: "*C*ardiac defects, *A*bnormal facies, *T*hymic hypoplasia, *C*left palate, and *H*ypocalcaemia."[61] Shprintzen rejected this and other designations that would replace his preferred name for the disorder, noting, "There is simply no valid evidence to suggest that velocardiofacial syndrome is aetiologically heterogeneous. The DiGeorge anomaly is known to be so, as is CHARGE [a related disorder]. Therefore, placing velocardiofacial syndrome, DiGeorge syndrome, and CHARGE under a single diagnostic category is an example of what used to be referred to as 'lumping,' which will only confuse clinicians, molecular geneticists, and most importantly, patients and their families." The DiGeorge "anomaly," Shprintzen argued, had multiple potential causes, whereas VCF syndrome had only one: the 22q11 deletion. This made it the one true syndrome. Developing a new big tent name for these disorders, he suggested, would represent inaccurate "lumping" by common clinical features, which medical genetics sought to avoid. Doing so, he argued, would only confuse physicians and patients who were already familiar with his term, *VCF syndrome*.[62]

Like Angelo DiGeorge, Shprintzen demurred when it was suggested that VCF syndrome should instead be named eponymously after him. Some authors did use the term *Shprintzen syndrome*, however, including David Smith in *Recognizable Patterns of Human Malformation*. Shprintzen made a practice of highlighting previous accounts of what he believed to be VCF syndrome, crediting the role of various researchers in its delineation.[63] In 1994 he wrote, "Many clinicians with different focuses of attention have described from a variety of perspectives what may be a single class of patients. If one believes that heart anomalies are the primary defect, DiGeorge syndrome becomes the diagnostic nosology of primary significance, whereas if one studies children with craniofacial anomalies, velocardiofacial syndrome may be of prominence." Shprintzen pointed to "the parable of the five blind men and the elephant," a metaphor that clinicians commonly used to describe situations in which researchers from various specialties observe the same clinical entity from different perspectives. In earlier accounts of what Shprintzen believed to be VCF syndrome, physicians from different specialties had primarily focused on immune, heart, or palate defects. He argued, however, that it took an awareness of patterns of malformation, coupled with the rigorous criteria applied in his clinic, for VCF syndrome to be recognized for what it was: a discrete but variable disorder.[64]

Shprintzen was not alone in citing the parable of the blind men and the elephant in reference to the 22q11 deletion. In 1996 Johns Hopkins human geneticist Eric Wulfsberg and colleagues published a paper titled "What's in a Name?," which considered the various disorders associated with the 22q11 deletion and used the parable as an epigraph. Providing a somewhat different interpretation of how the metaphor applied in this case, the authors noted, "Our analogy to this fable is not to imply some 'blindness' on the part of these clinicians, but rather to point out the well-known difficulty in delineating the indistinct phenotypic boundaries of a syndrome until a genetic or biochemical marker for the condition is available. The recent availability of a fluorescent in situ hybridization (FISH) probe to detect deletion of the DGCR [DiGeorge Chromosome Region] now allows delineation of the broad phenotype of our 'elephant.'" Rather than focusing on clinical experience, expertise, and rigorous criteria, as Shprintzen did, Wulfsberg pointed to the central importance of the molecular detection of the 22q11 deletion in demonstrating that DiGeorge and VCF syndrome were variable forms of one disorder. For these authors, it was the 22q11 deletion that allowed researchers to isolate the disorder in "pure culture." While clinical variability had previously been a problem for accurate diagnosis, they stated, "Since the correlation between the DGCR deletion and its phenotypic spectrum is rapidly and precisely being defined by hundreds of reported cases, we think this will diminish as an issue."[65]

By the mid-1990s medical geneticists widely accepted that, despite the phenotypic variability of DiGeorge and VCF syndrome patients, the common presence of a 22q11 deletion in each demonstrated that there was just one elephant in the room. Wulfsberg commented that what remained was "the obvious but difficult question of what to call the 'elephant.'" Despite initial enthusiasm, researchers soured to CATCH-22, because it was a literary term that had "negative" and "regrettable" implications. Wulfsberg dismissed an alternative term based on the causative mutation, suggesting that "the term 'del 22q11.2 syndrome,' while emotionally neutral, does not confer any phenotypic information." Instead, he promoted the compound term *DiGeorge/velocardiofacial* (DG/VCF) *syndrome* as "a reasonable compromise in referring to this condition, as it calls attention to the phenotypic spectrum using the most familiar historical name." Other influential participants in naming the "elephant" were not swayed by this compromise.[66]

The next year, Children's Hospital of Philadelphia clinicians Elaine Zackai and Donna McDonald-McGinn submitted "our elephant" to the *American Journal of Medical Genetics,* in the form of an illustration of the blind-men parable. The drawing depicted three men and three women examining an elephant from various perspectives and with differing instruments, which included a magnifying glass and a stethoscope. The elephant was wearing a banner with the number 22 on it, which represented the common chromosomal mutation. While this depiction presented "the

Drawing presenting the parable of the blind people and the elephant as it relates to DiGeorge and VCF syndromes. Each individual depicted is so focused on one portion of the elephant that each fails to see the "big picture." The elephant was numbered 22 to highlight the significance of the 22q11 deletion in demonstrating the existence of a single genetic disorder. *Source:* D. M. McDonald-McGinn, E. H. Zackai, and D. Low, "What's in a Name? The 22q11.2 Deletion," *American Journal of Medical Genetics* 72, no. 2 (1997): 247. Reprinted with permission from John Wiley and Sons and Donna McDonald-McGinn. Copyright 1997 Wiley-Liss, Inc.

big picture," a whole elephant marked by 22, each of the examiners was shown to be very focused on examining his or her own targeted interest.[67]

The Philadelphia group's illustration of the blind people and the elephant metaphor was a means of representing their own preferred term, which highlighted the central role of the 22q11 deletion in the cause and diagnosis of the disorder. As McDonald-McGinn and Zackai argued, focusing on the deletion "allows all physicians involved to immediately understand the cause of the problem, the recurrence risks, and the variable diagnosis. It also allows physicians to compare their patients to children in the literature with the exact same thing rather than possibly comparing 'apples to oranges.'"[68] More than just a diagnostic marker, the 22q11 deletion was presented as a genetic arbiter, which ensured that physicians were examining individuals in "pure culture" and describing a single elephant.[69] While everyone involved in the debate accepted this premise, some still rejected the suggestion of naming the elephant after the deletion. The disorder could be reduced to a single mutation, but the social interests associated with it could not.

An Institutionalized Syndrome

In 1997 Robert Shprintzen moved to the Upstate Medical University in Syracuse, New York, where he became the founder and director of the Velo-Cardio-Facial Syndrome International Center. He was also actively involved, and personally interested, in ongoing discussions about a universal name for disorders associated with the 22q11 deletion. The identification of this common mutation in patients with both DiGeorge and VCF syndrome proved to be a double-edged sword for Shprintzen. He pointed to the 22q11 deletion as definitive evidence that VCF syndrome was a distinct genetic disorder, which affected many patients who had been previously diagnosed with DiGeorge syndrome. At the same time, the common mutation had led medical geneticists to consider a new designation for the disorder, which would uproot his chosen name.

Throughout the first decade of the 2000s, Shprintzen continued to advocate for the ongoing use of the term VCF syndrome. In a 2008 book on the disorder, he responded to proponents of an increasingly common alternative name by noting, "Some authors have advocated for the use of 22q11.2 deletion syndrome. This seems curious for a microdeletion syndrome. . . . Microdeletion syndromes like VCFS are essentially never identified by their deletion site." Shprintzen then pointed specifically to Williams syndrome, which was associated with a deletion at 7q11 but had never been called 7q11.23 deletion syndrome.[70] His position would have held water five years earlier, but naming syndromes after microdeletions had become more common by 2008.[71] Having made this point, Shprintzen closed his statement with the comment, "We advocate calling the disorder VCFS, which is descriptive, geographically nonspecific, free of eponyms, and much easier to write and say than 22q11.2 deletion syndrome."[72] Each of these was a well-established argument for adopting a descriptive name, such as VCF syndrome, over other options. It must be acknowledged, however, that while VCF syndrome was not itself explicitly geographic or eponymous, the term was very much linked historically and institutionally to Shprintzen and his professional contributions, as well as to the institutions where he worked and that he had founded.

The primary advocates of the universal adoption of the alternative designation 22q11.2 deletion syndrome were Zackai, McDonald-McGinn, and Emanuel at the Children's Hospital of Philadelphia. They also institutionalized their preferred name in the *22q and You* Center and newsletter and in the International 22q11.2 Foundation, of which McDonald-McGinn was a founding board member. Through the foundation's "Same Name Campaign," they promoted a universal name. Their website claims, "The Foundation's position is clear: Rather than further dividing our small 22q community, we aim to unite and empower those affected by promoting use of the name currently recognized by the standard chromosome naming system: 22q11.2 deletion syndrome." These advocates acknowledged the difficulty for individuals in moving away from a lifelong identity and toward a new term. So they encouraged patients to use 22q11.2 deletion syndrome alongside terms like DiGeorge or VCF syndrome, if they preferred. Their long-term goal, however, was to encourage physicians to diagnose cases using only "22q11.2 deletion syndrome," so that "outdated names" would eventually disappear.[73]

Reflecting on why the term *VCF syndrome* remained in use over the years, while *DiGeorge syndrome* fell out of favor, Shprintzen noted, "I have personally been active in keeping the VCFS label visible because I believe chromosome location labels can be terribly inaccurate and lead to inappropriate assumptions about the syndrome. I know that clinicians and researchers in Philadelphia and at the Children's Hospital of Philadelphia at their 22q and You Center are pushing hard to have this called 22q11 deletion syndrome, but there are sound clinical and scientific reasons for rejecting that label."[74] Into the following decade, Shprintzen remained a driving force behind the continued use of the term he had coined four decades earlier. At this time, Shprintzen's opposition to the term *22q11.2 deletion syndrome* was aided by reports that not all deletions in the region resulted in VCF syndrome. Drawing on this new information, he noted, "I think that [genomic] location names are a problem, because there are deletions in 22q11 that don't cause the syndrome. So if the critical region is not involved, but other portions of 22q11 are, people don't have VCFS."[75]

Personal experiences with this condition expanded far beyond the institutions and people in New York and Philadelphia who disagreed about what it should be called. Affected families all over the world established their own identities around the disorder, and they were much more interested in receiving treatment and support than in debates over what to call themselves.[76] Other support centers moved in the direction of a compound or broadly inclusive name, along the lines of 22q11.2 deletion/VCF syndrome, while those interested in maintaining a single name for the disorder remained entrenched. Robert Marion, a student of Shprintzen's, wrote in a 2014 essay, "Whenever I encounter a patient with the constellations of findings that are caused by a deletion of the q11.2 segment of chromosome 22, I will refer to that patient's conditions not as DiGeorge syndrome . . . nor as 22q11.2 Deletion syndrome; instead, I will always call the condition 'velo-cardio-facial syndrome.' Thank you, Bob!"[77] The institutionalization of these terms continued through those who had trained in centers that supported one or the other.

Throughout his career, Shprintzen built social institutions and a professional following around VCF syndrome. Though he retired from clinical practice in 2012, he remained active in the VCF community as a supporter of hundreds of patients around the country. Unmoored from the brick and mortar institutions that had sustained his career for decades, Shprintzen moved his advocacy online through a new project, The Virtual Center for VCF Syndrome. Rather than bringing the world's VCF syndrome experts together under one roof, Shprintzen developed an online support system for patients, which included his many colleagues and supporters around the world. While the group provided advice to patients who might identify by various syndrome names and was willing to refer families to institutions that went by different terms, it continued to do so under the designation that Shprintzen coined and promoted for decades: VCF syndrome.

Conclusions

The life histories of genetic disorders in postwar medical genetics were malleable, rather than fixed, narratives. As first described in the 1960s and 1970s, DiGeorge and VCF syndromes were incommensurable. Physicians expected DiGeorge's pattern of malformation to be lethal long before the developmental pattern of VCF syndrome could be clinically recognized. By the 1990s, following decades of extensive clinical and genetic observation, medical geneticists broadly agreed that these two disorders should be associated with the same address in the chromosomal infrastructure for diagnosis and prevention. However, as described in this chapter, the identification of a common mutation in the early 1990s did not constitute the beginning or the end of the integration of these disorders' life histories. Even with the 22q11 deletion in place as a definitive means of delineation, many clinicians pushed back against the idea that this mutation alone should be used to characterize the disorder. The 22q11 deletion did not erase the social identities and institutions that had already been built around these disorders. Biosociality was not automatically realigned to reflect the identification of a common mutation.[78]

I argue that the one mutation–one disorder ideal of medical genetics was strengthened in part by its self-reinforcing qualities. Indeed, what counted as a disorder or as a mutation was often simultaneously constructed by medical geneticists, as they worked to characterize groups of patients in the laboratory and the clinic. In many instances, including the representative historical study presented in this chapter, the balance of the one mutation–one disorder ideal was heavily weighted toward the side of mutations as definitive indicators of distinct disorders. DiGeorge syndrome was initially delineated based on a similar pattern of malformation. However, the suspicion of genetic heterogeneity, along with the subsequent identification of deletions at 10p13 and other chromosomal addresses in some patients, was sufficient to overrule the significance of clinical similarity for medical geneticists. Many instead began to focus on the 22q11 deletion as the most reliable indicator of a distinct disorder.

Clinical features, however, did still hold weight in postwar medical genetics. As described in chapters 3 and 4, patients with Angelman syndrome were found in the late 1980s to have the same mutation that had been associated with Prader-Willi syndrome. Because of the distinct clinical features of these disorders, however, medical geneticists thought it impossible that the same genetic mutation caused them both. In this case, a common mutation was not the sole arbiter of syndrome delineation. Rather, the implausibility that Prader-Willi and Angelman constituted a single genetic disorder pushed medical geneticists toward alternative explanations. The life histories of DiGeorge and VCF syndromes provide another example of how medical geneticists' applied the one mutation–one disorder ideal. In the early 1990s, a common mutation became the corroborating evidence that definitively proved what many medical geneticists had already assumed based on the similar clinical features of two historically distinct populations. It was determined that individuals who shared the 22q11 deletion had the same disorder.

Looking beyond considerations of accurate delineation and naming, an important question remained for medical geneticists: what value did patients and families gain from the association of their conditions with specific genetic mutations? Sociologist Daniel Navon has described instances in which the identification of a common mutation among individuals who would not have been grouped together based on similar clinical features offered families a social and medical identity, and a basis for alliance, that was never before possible.[79] The identification of a mutation promised improved diagnosis and inclusion in medical geneticists' chromosomal infrastructure, as well as the potential hope for gene-specific therapy. However, treatment options, beyond the potential for prevention through abortion, rarely materialized. Indeed, as historians Soraya de Chadarevian and Harmke Kamminga incisively noted, "Molecularization has rarely converged successfully on etiology, diagnosis, and therapy conjointly."[80]

Reflecting on the actual products of decades-long efforts by medical genetics to identify disease-causing mutations, the clinician and informally trained dysmorphologist Robert Shprintzen lamented, "[Molecular genetics] is disassociating the disease from the patients." Looking back to earlier approaches, prior to the rise of molecular cytogenetics in clinical diagnosis, Shprintzen noted, "That is what David Smith did very well; he tied everything to the patient, not their gene."[81] While the 22q11 deletion had aided the advancement of his own career in the 1990s, in retrospect Shprintzen acknowledged that, for patients and families, molecular cytogenetics research and designations were only as valuable as the treatments they produced.

In the early twenty-first century, as the molecular approaches of medical geneticists generated increasingly high-resolution findings, chromosomal addresses eventually proved to be too blunt a designation for isolating many disorders in "pure culture." This included DiGeorge/VCF syndrome. In 2014 medical geneticists reported that another clinically distinct disorder was also caused by a different 22q11.2 microdeletion.[82] Nonetheless, as described in the next chapter, even as the density of molecular analysis increasingly made chromosomal designations seem old-fashioned, medical geneticists continued to depict, conceptualize, and discuss the human genome and its morbid anatomy at the level of banded chromosomal ideograms.

Chapter 6

Getting the Whole Picture

In 2011, during the early days of my research for this book, I interviewed Beverly Emanuel, a medical geneticist at the University of Pennsylvania and the Children's Hospital of Philadelphia. When our conversation was drawing to a close (and after I had already turned off my tape recorder), Emanuel was reflecting on her research career, which had largely been defined by her focus on a particular subregion on the long arm of chromosome 22. As Emanuel mused, she held her thumb and first finger about a centimeter apart and spoke of how amazing it was that something so small could be the focus of an entire (quite productive and rewarding) career. Now, she knew as well as I did that the genomic region she worked on is much smaller than the space between her fingers; in fact, it is almost imperceptibly small. But that bit of space between her fingers was real to her in terms of how she thought of chromosome 22. Visualized as an ideogram, the chromosome was a tangible entity, maybe a few inches in length. Small portions of this chromosome have been the focus of many a life's work.

Many of the genomic entities and processes that Emanuel and thousands of other postwar medical geneticists had dedicated their lives to studying were too small to ever be directly made visible. However, the work objects of these researchers were far from invisible to them. In line with generations of biomedical researchers before them, these individuals relied on painstaking observation and standardized ways of seeing and communicating, in practicing their trade. What was the point of retaining chromosome-level conceptions and depictions of the human genome in an era when clinicians and researchers had a complete DNA reference sequence at their fingertips? Why had an "antique" nomenclature, which was developed in the 1970s for the description of chromosomes, remained a prominent set of landmarks in the post–Human Genome Era?[1] The answers to these questions are particularly perplexing when one considers that chromosomal banding nomenclature and genomic sequence data have always been incommensurable: banding boundaries could, at best, be located within a range of 100,000 DNA base pairs. This chapter considers the forces that have helped to maintain the importance of chromosome-level nomenclature and reflects on why older languages of description based on cytogenetic analysis remain intact despite the presence of newer and more exacting options. Here I explore the continued use of chromosomal depictions and addresses in medical genetics communication both among physicians and by physicians in consultation with families. I argue that chromosomes were retained in the post–Human Genome Project era because they offered tangible and familiar organizational referents in a time of increasingly abstract genomic analysis.

Beginning the 1990s, new genetic testing approaches were developed that were designed for use in molecular-level whole genome analysis, including comparative genomic hybridization and DNA microarray. Medical geneticists adopted these techniques to screen the human genome for tiny variations and mutations at specific chromosomal locations. Relying on the information made available by the still growing chromosomal infrastructure for diagnosis, medical geneticists correlated many new instances of genomic variation and mutation with particular patterns of malformation. Applied in the context of prenatal diagnosis, these techniques, which in the earliest years of the 2000s increasingly became commercial products, promised to provide physicians and parents with "the whole picture." The marketability of these genome-wide untargeted scans was rooted in growing expectations among physicians and consumers that all pregnancies were at risk for genetic disease.

This chapter begins in the 1990s, as researchers sought to enhance the power of medical genetics by shifting from disease-specific FISH probing to whole genome analysis. As the Human Genome Project neared completion at the end of the decade, molecular cytogeneticists were engaged in an extensive project using FISH probes to bridge molecular and cytogenetic maps of the genome. They sought to link findings from whole genome molecular screening to the more familiar and long-standing chromosomal infrastructure for diagnosis in medical genetics. During the early years of the 2000s, medical geneticists used whole genome screening techniques to search for chromosomal mutations that were considered too small to be made visible under the microscope. Many new variants were identified in patients affected by disorders of unknown cause, and medical geneticists struggled to determine which of these variants had pathological impacts. I examine the interpretation and communication of results from microarray and the implications of its application in the prenatal context. I continue to explore how new mutations were identified and stabilized in the laboratory and the clinic and how they were put to use for predictive diagnosis and prenatal prevention.

From Targeted FISH to Whole Genome Analysis

Throughout the 1990s, researchers in the emerging hybrid area of molecular cytogenetics sought to identify and make microscopically visible molecular probes for specific genomic locations. Various techniques based on radioactive and fluorescent detection were developed to this end. As described in chapter 4, one particularly successful approach was fluorescence in situ hybridization (FISH), which was developed during the late 1980s, broadly taken up in the 1990s, and has continued in widespread use. FISH probes were adopted to test for the presence or absence of targeted chromosomal segments. This analysis could be preformed without making chromosomes directly visible, which was one of the most difficult and time-consuming aspects of cytogenetics. Because of the ease and efficiency FISH brought to analysis, medical geneticists adopted it for identifying the common trisomies (13, 18, 21) and to test for the presence or absence of smaller chromosomal regions associated with disorders such as DiGeorge syndrome. Notably, the use of FISH was targeted to testing for, or verifying the diagnosis of, specific disorders. A FISH test was pursued in cases where a physician already believed, based on clinical features, family history, or other risk factors, that a certain mutation might be present.

FISH differed from traditional karyotype analysis of all the chromosomes at once, which had for decades offered a valuable "glimpse" of the entire human genome and had occasionally turned up chromosomal abnormalities that had not been anticipated but were clinically meaningful. Indeed, traditional human cytogenetic analysis represented the earliest form of whole genome analysis.[2] Molecular probing techniques such as FISH allowed for high-resolution analysis and the detection of submicroscopic mutations, but in practice they represented a limited form of hypothesis testing. Clinicians had to know what they were looking for to order a targeted assay. FISH was not going to generate new, unanticipated or unpredictable results.

Responding to these limitations, during the early 1990s some of the same molecular cytogeneticists who had developed FISH began to work on new approaches that would combine the breadth of traditional chromosomal analysis with the depth of molecular probing. Once again, Joe W. Gray's laboratory at the University of California, San Francisco, was a leader in the development of this new technique, called comparative genomic hybridization (CGH). Clinically trained Finnish researchers Anne and Olli Kallioniemi led this project, which aimed to provide a broad molecular cytogenetic analysis of tumor cells. In their initial 1992 report of the technique, the authors identified the strengths and weaknesses of existing cytogenetic and molecular approaches and highlighted the novel capabilities of CGH, noting, "Molecular methods are highly focused; they target one specific gene or chromosome region at a time and leave the majority of the genome unexamined. We have developed a molecular cytogenetic method, comparative genomic hybridization (CGH), that is capable of detecting and mapping relative DNA sequence copy number between genomes."[3] Rather than focusing on just one small region at a time, CGH allowed for the large-scale comparison of entire human genomes at the molecular level.

The Kallioniemis used CGH to compare a cell line from a breast cancer patient with a reference genome from a clinically normal female. The genome from each cell line was cut into millions of DNA fragments and fluorescently labeled, the cancer genome in green and the reference in red. The labeled genomes were then allowed to competitively hybridize in situ with normal chromosomes. In most locations both the red and the green DNA annealed to the chromosomes, producing a combined yellow signal. Some locations appeared either green or red under the microscope. In green locations, the researchers inferred, more of the cancer patient's DNA had hybridized, indicating a duplication of genetic material in this area of the genome. Red signal, on the other hand, identified locations where DNA was deleted in the genome of the breast cancer cell line.[4] Denis Rutovitz, a collaborating author at the University of Edinburgh, developed image processing software for use with CGH; it facilitated the genome-wide examination of green and red signals, allowing for the chromosome site-specific quantification of each. CGH analysis of various cancer cell lines revealed numerous duplications and deletions of genetic material on chromosomes. These were interpreted as corresponding to the "amplification" (presence of one or more extra copies) of specific genes involved in causing cancer.[5]

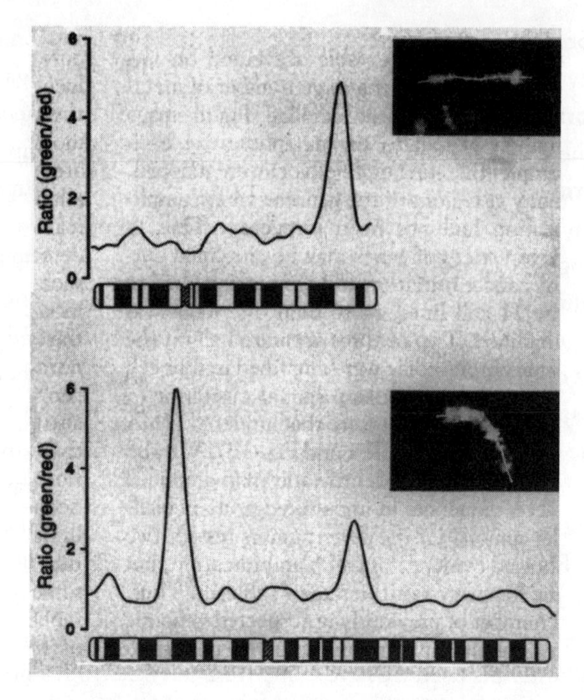

Demonstration of the use of array-CGH to identify the amplification of genetic material in small segments of chromosome 8 (*A*) and chromosome 2 (*B*). *Source:* A. Kallioniemi et al., "Comparative Genomic Hybridization for Molecular Cytogenetic Analysis of Solid Tumors," *Science* 258, no. 5083 (1992): 819. Reprinted with permission from AAAS.

Researchers in Gray's laboratory went on to examine many different cancer cell lines with the CGH technique. Olli Kallioniemi wrote in a 1993 paper, "The obvious advantage of CGH over previous molecular genetic methods is that amplification anywhere in the genome can be detected. No specific probes for or prior information of the amplified loci are required. . . . CGH also accurately maps the amplified regions to normal chromosomes."[6] In effect, CGH was the first molecular approach for genome-wide analysis that could be used to locate new chromosomal abnormalities associated with cancer and other disorders. CGH was in situ hybridization in which the entire genome served as a set of probes, instead of a specific segment.

During the mid-1990s, molecular cytogeneticists primarily applied CGH to studying cancer cell lines, which often had multiple significant chromosomal abnormalities. This included the presence of "marker" chromosomes, which were small duplicated portions of larger chromosomes that were often difficult to identify based on their banding pattern. CGH was particularly useful for identifying the origin of these marker chromosomes and other instances of amplification and deletion in specific genomic regions. When multiple cell lines with the same type of cancer were examined, researchers identified locations of reoccurring abnormalities, which pointed to the genomic location of potential cancer genes.[7] CGH was particularly valuable for medical geneticists in characterizing complex karyotypes in cancer patients, but it was initially less useful for identifying the subtle mutations that caused most genetic disorders.

DNA Chips and Arrays

While in situ CGH assisted medical geneticists in identifying chromosomal changes that they might have overlooked microscopically or had been unable to characterize, the technique did little to improve the resolving power of cytogenetic analysis. This was because researchers still had to microscopically detect signals along chromosomes to determine their genomic location. In the mid-1990s, compacted chromosomes in a human cell remained the most directly accessible and complete source of genomic DNA for hybridization. Molecular cytogeneticists estimated that amplified regions had to be at least 2 million base pairs long to be made visible using in situ CGH, and deletions would have to cover 10 million nucleotides to be visible.[8] In response to these limitations, molecular cytogeneticists began to develop new methods for high-density hybridization analysis that relied upon the use of DNA "chips" or "arrays."

Rather than hybridize to physical chromosomes, molecular cytogeneticists wanted to use short, chemically synthesized DNA fragments, oligonucleotides, as an alternative template for hybridization. Various methods had been developed to construct and spot these oligonucleotides onto a tiny DNA chip, which was a 1.28 cm square. From here, the efficiency of hybridization with genomic DNA from various sources could be examined.[9] The Oxford University laboratory of Edwin Southern was active in developing such methods, as was the Santa Clara, California, biotechnology company Affymetrix, which was founded by biochemist Stephen Fodor in 1994 to develop and market DNA chip technology. During the mid-1990s, Affymetrix demonstrated that DNA chip technology could be used to sequence small genomes, such as that of human mitochondria, and to identify variations within them.[10] DNA chips represented a shift away from in situ techniques, which relied on visible chromosomes as standards for comparison, and toward large-scale in vitro hybridization and computer analysis.

In the late 1990s, molecular cytogeneticists believed that applying DNA chip technology to the high-resolution analysis of the entire human genome would require significant scaling up of chip capacity and knowledge of the chromosomal location of many more unique DNA sequences. A group led by Peter Lichter in Heidelberg, Germany, sought to increase the resolving power of CGH using DNA arrays. In 1997 they reported on the development of a DNA array on a microscope slide, which allowed for the identification of genomic deletions and amplifications in multiple cancer cell lines. Instead of synthesizing their own oligonucleotides, the group drew on DNA libraries of specific chromosomes to build their array. Chemical techniques were used to "spot" specific DNA fragments to discrete locations on an array slide. Lichter and colleagues ultimately used their platform to identify new microdeletions in cancer cells as small as 75 thousand base pairs. Once fragments that were deleted or amplified had been identified, FISH analysis was used to determine the exact chromosomal address of the abnormality.[11]

The next year, Joe Gray, David Pinkel, and colleagues in the United States reported on their own "array CGH" technique, which they believed could be scaled up to whole genome analysis. Their proof of principle was performed using DNA fragments from libraries of chromosome 20, which were spotted onto an array. The researchers used this setup to identify specific genetic abnormalities in a breast cancer cell line. Normally, cells contain two copies of every chromosome and its components (genes and the nucleotide sequences between). Pinkel and Gray's report highlighted the identification of small deletions and amplifications of chromosomal material. These "copy number variants" were linked to specific cloned fragments of DNA, which were associated with specific chromosomal addresses by FISH analysis. Each was used as a starting point for further research on a specific genomic region.[12]

At the time of Gray and Pinkel's report, the Human Genome Project (HGP) was already in its late stages. Though it would provide a sequence-level molecular map of the genome, it was clear to geneticists that completion of the HGP did not mean the end of cytogenetic analysis; far from it.

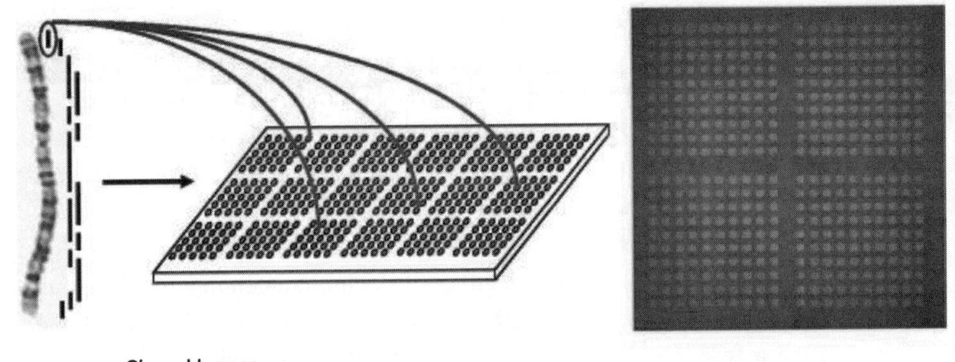

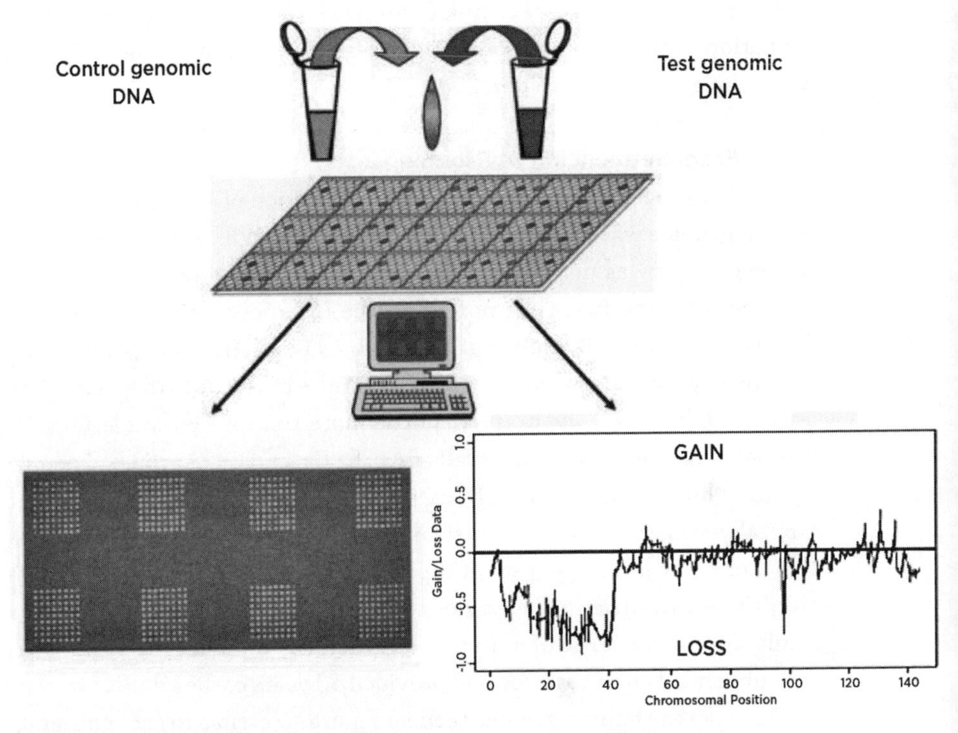

Demonstration of the process behind array-CGH. Control and test DNA were both allowed to hybridize with the array platform. A computer program measured the relative hybridization of each to identify deletions and duplications of genetic material along each human chromosome. *Source:* L. G. Shaffer and B. A. Bejjani, "A Cytogeneticist's Perspective on Genomic Microarrays," *Human Reproductive Update* 10, no. 3 (2004): 223. Reprinted with permission from Oxford University Press.

Indeed, there was a strong sense throughout the 1990s and into the following decade that human genome maps at various levels, from visible chromosomal banding down to nucleotide sequences, would need to be linked in order to achieve the many clinically related promises of the HGP.[13] As medical geneticist Julie Korenberg put it in 1999, "Translating problems of human disease into the language of the human genome requires a unified resource that bridges DNA sequence through chromosome bands. Such a resource must link . . . chromosome bands visible in single cells, and ordered clone arrays."[14] Even with the complete human genome sequence at their fingertips, researchers would need bridges back to chromosome-level mutations and analysis. Ultimately, a large-scale analysis of FISH probes would be necessary to link these two levels of genome mapping information. This project was already under way throughout the HGP and was completed along with it in 2000.

From Sequencing to Browsing

The official announcement of the completion of a rough draft of the human genome was made on June 26, 2000, at the White House. The DNA sequence itself was not made publicly available until July 7, when the genome was for the first time posted on the Internet by the University of Santa Cruz Genome Bioinformatics Group. As David Haussler put it, "That was the day that the world got the first glimpse of the human genome." In July 2000, however, the draft was little more than 2.7 billion letters: "It was nothing more than a waterfall of As, Ts, Cs, and Gs. So you had people counting how many times GATTACA appeared in it, or looking for secret biblical messages. . . . It was something you could use for wallpaper." While this molecular-level "glimpse" of the human genome was new and impressive, there was little in the way of contextual information or landmarks available to make sense of it. In this regard, it was similar to the "glimpse" that unbanded chromosomes had provided 50 years earlier.[15]

To make the human genome sequence more accessible to the thousands of clinicians and geneticists who expected to begin using it for diagnosis and research, multiple genome "browsers" were built during the second half of 2000. Included were the Map Viewer created by the National Center for Biotechnology Information (NCBI) at the National Institutes of Health in the United States; the Ensembl Genome Browser, sponsored by the Wellcome Trust Sanger Institute and the European Molecular Biology Laboratory; and the University of California, Santa Cruz (UCSC), Genome Browser.[16] These browsers served as portals to the raw human genome sequence data and provided the online software and annotations necessary to make the information useful. Haussler at UCSC said of the human genome, "In terms of usefulness it wasn't until this browser was built that people could actually use it."[17]

Genome browsers made exploring and analyzing the human genome a lot like browsing a bookshelf. The developers of these data portals sought to provide a top-down view of the genome, situating the human chromosomes as the primary organizational units. One could jump straight to a particular gene or genomic region, or begin with a specific chromosome and zoom in from there. When I interviewed Haussler in his office at UCSC, he immediately asked for my favorite gene, so that he could search for it in the Genome Browser. I picked SNRPN, a gene associated with Prader-Willi syndrome. Haussler typed this into the search mechanism, and the browser brought us to a region near the centromere on the long arm of chromosome 15.

The UCSC Genome Browser had a horizontal orientation, with a series of customizable data tracks appearing on the screen. In its default mode, one was shown the nucleotide number of the region in question (the DNA nucleotides on each chromosome were numbered from 1 into the hundreds of millions, beginning at the farthest point from the centromere on the short arm). Below this, the expanse of the gene and its coding regions were shown, and further down were additional data tracks, including the sequence homologies with various other organisms. Above all of this information, featured prominently at the top of the page, was a banded ideogram of the human chromosome being explored; in the case of SNRPN, it was chromosome 15. A red box (or line, depending on how zoomed-in the tracks were) showed the location and extent of chromosome 15 that was currently being viewed. SNRPN fell into the chromosomal band 15q11.2 and appeared to be quite close to the boundary with the next visible band, 15q12.

Chromosomal ideograms had a similarly prominent position in the other major genome browsers. In Ensembl, one began with a vertically oriented view of all 24 human chromosomes, the "whole genome," and from there could click on a particular band, or drag the cursor and select a larger region of one chromosome. Otherwise, Ensembl offered an experience similar

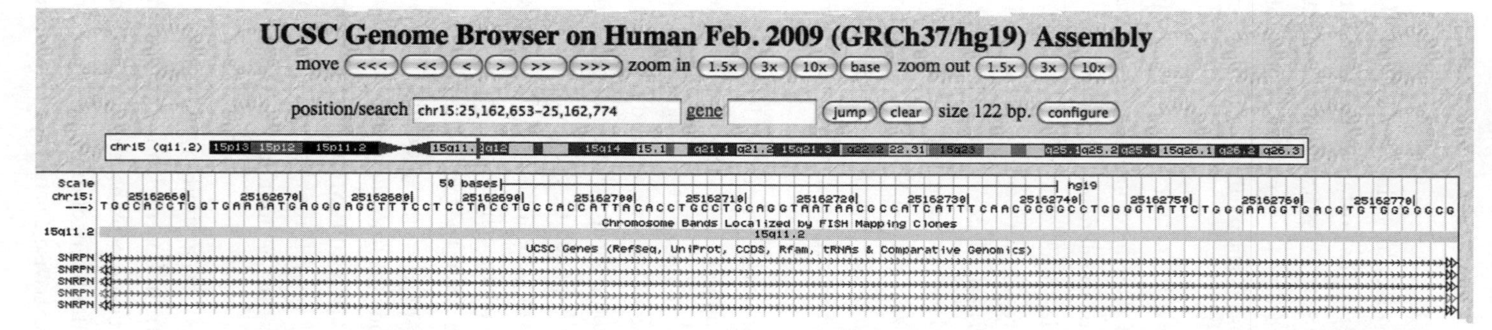

Web-based readout from the University of California, Santa Cruz, Genome Browser. A segment of human genomic DNA sequence is shown from a small portion of the chromosomal band 15q11.2. The banded ideogram for chromosome 15 is depicted for reference. The vertical bar through the ideogram at 15q11.2 indicates the approximate location of the DNA sequence shown below. *Source:* University of California, Santa Cruz, Genome Browser, Assembly GRCh37/hg19, http://genome.ucsc.edu/.

to that described with the UCSC Genome Browser. A horizontal ideogram of the chromosome being examined was shown above and a series of data tracks below. The NCBI Map Viewer also had several common experiential features. For instance, one began by choosing a chromosome from a vertical depiction of 24 unbanded chromosomes. The interface was noticeably different at the chromosome level, however, because it was vertically oriented, with a chromosomal ideogram on the far left and a series of data tracks to the right. The NCBI depiction of chromosomes was more in line with early genomic representations, which depicted chromosomes vertically with their short arm on top. In the era of widescreen computers, horizontal depictions of ideograms became standard.

Initially, UCSC and NCBI provided their own, slightly different assemblies of the human genome sequence. However, human and medical geneticists eventually decided that, for clarity, there should be one reference genome sequence shared by all browsers. The NCBI assembly was chosen.[18] Each browser continued to have its own annotational and organizational strengths. The Ensembl browser specialized in highlighting protein structure and function, while the UCSC Genome Browser was more focused on the genetic code itself. Human geneticists therefore often went to the USCS browser first, and then jumped into the Ensembl browser for protein analysis. The NCBI Map Viewer was particularly closely integrated with the Online Mendelian Inheritance in Man database and excelled in highlighting genes and their chromosomal positions.[19]

The official publication of the draft genome sequence and related reports came in issues of *Science* and *Nature* in February 2001. By agreement, papers concerning the results from the publicly funded Human Genome Project appeared in *Nature,* and those from Craig Venter's competing private venture, which also finished its draft in June 2000, were published in *Science.* The primary collaborative paper from the public project in *Nature* highlighted the newly constructed UCSC and Ensembl genome browsers.[20] Published alongside a series of related papers on different aspects of the HGP and its preliminary results was a foldout map depicting the genome at the visible level of chromosomes. A vertical, microscopic image of each chromosome was placed next to a series of horizontal data tracks. Like the genome browsers, these tracks broke down each chromosome into cytogenetic banding units and nucleotide base pair distances. Certain genes were also listed along each chromosome.

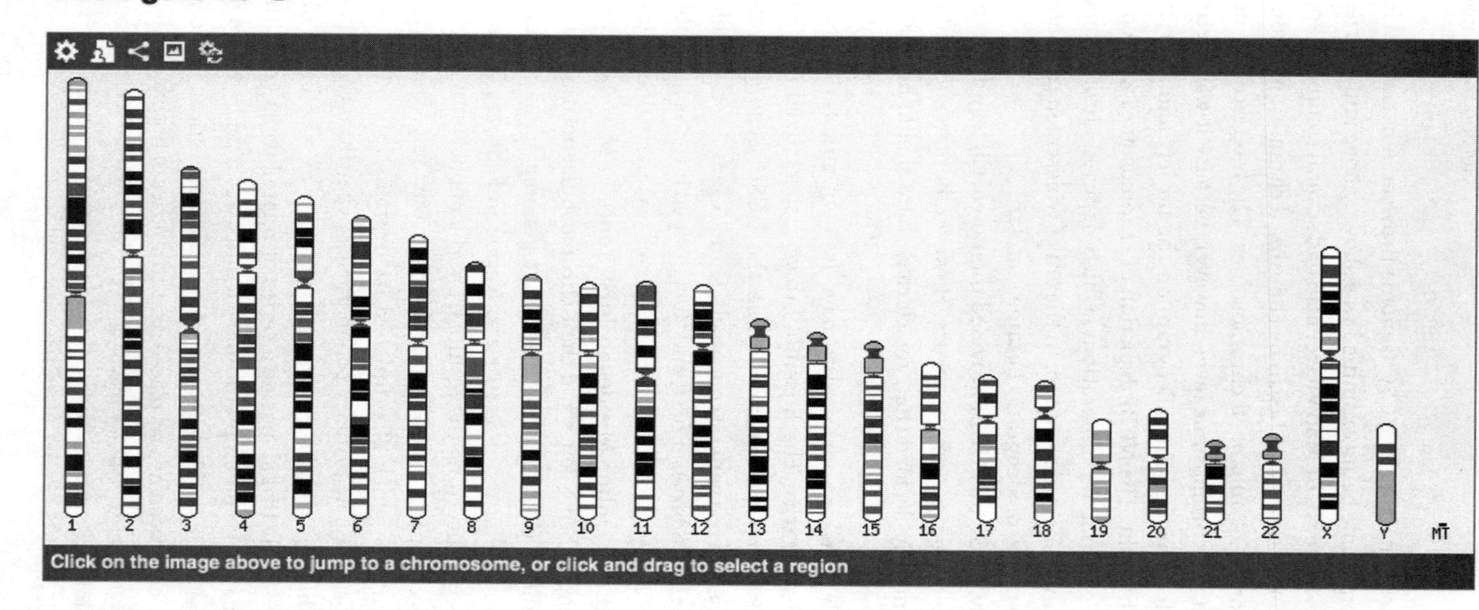

Depiction of the "whole genome" made up of banded ideograms, from Ensembl Browser. Clicking on a specific band allowed the user to browse the genes in that portion of the human genome. *Source:* Ensembl Genome Browser, www.ensembl.org/Homo_sapiens/.

When I walked into Haussler's office, I noticed that he had the *Nature* foldout displayed prominently on his wall. I asked him why the contributors to the Human Genome Project's primary publication had decided to present this large mass of molecular data, which they had spent years sequencing and compiling, in a traditional chromosomal arrangement. He responded, "People needed to have something tangible; they wanted a foldout. It was a monumental achievement, so you wanted something you could physically touch and look at to get an idea of the scope of the work. Previously you would publish a paper in *Nature* based on five years work of locating one of those genes, and Figure 1 would be the chromosomal ideogram and your location of your gene. And that was an incredible achievement, and I think this made a statement of, wow, look at the scale, all at once, all those genes."[21] After a decade of deciphering all 3 billion nucleotides that make up the human genome, it remained the case that the best way to capture and make apparent the broad implications of this great accomplishment was to visually depict it at a much lower level of resolution. Even on this poster-size foldout figure, each linear inch represented 10 million DNA base pairs. In this sense, the figure certainly conveyed the immensity of the data set that had been obtained.

Published along with the initial HGP publications in early 2001, this foldout represented a significant continuity in conceptions of the genome before and after the completion of a draft sequence. Clinical and research understandings of the genome as a visible, tangible entity did not immediately fade away or become irrelevant. Rather, these conceptions remained central to thinking and practices in the post-HGP era. This continuity was particularly striking to me during the interviews I conducted with medical geneticists.

Post-HGP Chromosomal Thinking

One of the articles in *Nature,* published alongside the initial paper on the completed human genome draft sequence, reported on a large project to integrate cytogenetic landmarks into the genomic DNA sequence. Molecular cytogeneticists had used 7,600 FISH probes derived from DNA fragment libraries to correlate the chromosomal banding nomenclature with the draft sequence of the genome. Each probe was fluorescently tagged, facilitating the determination of the chromosomal band where it hybridized in situ, along with the relative ordering of probes that localized to the same chromosomal banding address. The probes each contained a known DNA sequence, which offered a bridge back to a location at the nucleotide level. The project's goal was to correlate two existing maps of the genome: one based on observational analysis of the human chromosomes going back to the 1970s, and the newly completed draft DNA sequence.[22]

To improve the accuracy of the genome browsers, molecular cytogeneticists wanted to know where the boundaries of the consecutive chromosomal bands fell within the genome's sequence. Realistically, however, these data sets were incommensurable. How closely could the banding boundaries be associated with the reference sequence? Haussler noted that estimations were within "100,000 bases at best, and that's assuming fairly dense mapping, in the optimum conditions. In sparse places, it is a million bases." Where one visible band ended and the next began had to do with the physical compaction of chromosomes, which was indirectly related to the DNA code (bands with active genes were less compacted and appeared lighter).[23]

Nonetheless, the approximate correlation between chromosomally visible bands and the genomic code was of value in genetic research and gene mapping. As the authors of the *Nature* FISH analysis report explained, "To proceed from cytogenetic observation to gene discovery and mechanistic explanation, scientists will need access to a resource of experimental reagents that effectively integrates the cytogenetic and sequence maps of the human genome."[24] Often (as is seen throughout this book), the first indicator of the genomic location of a disorder came from the identification of a chromosomal abnormality in similarly affected individuals. A better correlation between chromosomal addresses and genomic sequences locations helped medical geneticists in targeting their search for disease-causing genes. UCSC researcher Terrence Furey, in his 2003 report of another study, involving 9000 FISH probes, similarly argued that "the integration of the cytogenetic map with this draft sequence provided cytogeneticists with the necessary link to this molecular-based resource. Given a chromosomal abnormality in a diseased cell where the affected region has been cytogenetically mapped, the corresponding area in the draft sequence can be easily determined, and then investigated for potential disease genes."[25]

Efforts to correlate the cytogenetic banding map with the draft human genome sequence were about more than targeting the search for new disease genes. From a conceptual perspective, the chromosomal banding map helped medical geneticists to find their way in the genome and to communicate genomic locations to their patients and colleagues. As David Ledbetter put it, "It's hard to talk about a gene using genome sequence coordinates, because how do you visualize that, how do you wrap your brain around genome sequence coordinates?"[26] Ledbetter was referring to the system by which each nucleotide in the human genome reference sequence was numbered on each chromosome from 1 into the hundreds of millions. For instance, one could point to the gene SNRPN as being located on chromosome 15 between the nucleotide coordinates 25,217,650 and 25,224,945, or as being at the chromosomal address 15q11.2. The first method was more exact but did not provide a familiar sense of location within the human genome or on chromosome 15.

Other physicians and geneticists offered similar accounts of how their conceptions of the genome remain grounded in human chromosomes. As Beverly Emanuel expressed it, "If you go to any of the sequence websites, like the [UCSC] Genome Browser, and you focus in, there is an ideogram that is still there . . . because for so long we have used that information in that way, and it does help to put that in a perspective, as opposed to a long string of numbers. A long string of numbers, from 17 to 20 million, doesn't necessarily put you into a visual of where in the genome it is."[27] Chromosome bands offered a better sense of location than nucleotide numbers did. In addition, banding nomenclature locations helped to historicize a disorder and its genomic association. Medical geneticist Kurt Hirschhorn described the importance of chromosomal context in this way, once again referencing the UCSC Genome Browser: "Some of this is historical: a number of difficulties have been described by virtue of a chromosome and a position in a chromosome. So, if you want to understand what the background of the whole thing is, you really need to see the chromosome. And I think they have done a very good job of that at Santa Cruz."[28]

Chromosomal locations were also more useful than genomic coordinates when it came to communicating with patients. As medical geneticist Uta Francke noted, "It's hard to visualize just DNA. . . . If you talk to parents, or people who are affected with a chromosomal imbalance, it helps a lot to show them a picture of a chromosome, and say look, this is the piece that is now translocated. It gives them some coordinate numbers."[29] Indeed, it was much easier to offer patients a visual representation of a genomic abnormality than to give a sequence-level account. Obstetrician and clinical geneticist Deborah Driscoll offered a similar narrative of how she used chromosome-level explanations to counsel patients, "We have come a long way [with molecular genetics], and I think it has really changed the way we can counsel families, but visually you still think of a chromosome. It is kind of where the DNA lives, the genes live, so that's how I think of it. When I talk to patients, I talk about chromosomes, and then what I try to do is explain to them what is a gene."[30]

Many medical geneticists noted that those who regularly worked with the human genome had a common sense of its geography based on the chromosomal banding nomenclature developed in the 1970s. Indeed, one could not know the entire human genome at the level of resolution that the DNA reference sequence provided. But medical geneticists were familiar with conceptualizing the genome as a karyotype. Referring to Victor McKusick's morbid anatomy of the human genome, his former student Reed Pyeritz suggested, "I think if you had psychoanalyzed people back then and tried to get them to express what image flashed in their mind when they thought of the genome that [McKusick's chromosome level gene and disease maps] would be it. . . . I still have tucked away the notion of the karyotype . . . but now I think of a cloud, it's just a mass of data."[31] In the era of whole genome sequencing, it remained difficult for medical geneticists to conceive of the genome without referring to chromosomal locations and depictions.

While the human genome reference sequence was an impossibly large and repetitive data set, chromosomal nomenclature offered researchers and clinicians a satisfying sense of place. Medical geneticist Dorothy Warburton suggested that she thought of the human genome as a familiar neighborhood, full of landmarks that make navigating it easy and intuitive. To remove these landmarks and replace them only with consecutive numbering, she noted, would be to take away the native elements that make a neighborhood recognizable.[32] Ledbetter expressed a similar feeling, that chromosomal locations offered him a sense of place and context:

> If I am in a seminar or talking with somebody and they start talking about a gene, the first question I ask is, what chromosome is it on, where does it live? And, it is sort of like saying, where are you from? Just the geography of where

> somebody lives or comes from just helps you. . . . If a gene is on chromosome 18 or it is on chromosome 16, I'm not really asking because I want to know what the individual gene neighbors are. I just can't imagine a gene without thinking where it is in the genome.[33]

As Ledbetter went on to explain, there was not an explicit functional purpose for knowing on which chromosome a gene was located. Indeed, genomic (i.e., chromosomal) proximity did not suggest that there was a functional relationship or interaction among gene functions or products.

Knowing where a gene was located in the human genome's morbid anatomy also offered researchers important context in terms of what other genetic entities and regulatory elements were in the area. This information could be relevant in cases involving chromosomal aberration or genomic imprinting (as in Prader-Willi and Angelman syndromes). Molecular geneticist Robert Nicholls said, "I'm interested in where genes are, or abnormalities are in terms of a chromosome simply because it is an amazing guidepost to what genes are there, what microRNAs are there, that type of thing, so for me the chromosomal correlation is a guidepost as to the underlying genes or regulatory sequences."[34] Driscoll similarly referred to cytogenetic banding nomenclature as "the signposts along the genome."[35] While many new signposts had been developed, chromosomal addresses continued to provide a broadly shared, visual language for describing familiar locations within the genome. Medical geneticists worldwide were already familiar with chromosomal landmarks, and this made them highly useful for communication.

Although the human genome was increasingly conceptualized and presented as an expansive cloud of data in the post-HGP era, medical geneticists continued to rely on familiar landmarks and low-resolution chromosomal representations of the human genome as they interacted with it. Chromosomal ideograms offered a tangible landscape within which medical geneticists situated and contextualized their research, and they also offered a visual reference for use in counseling patients on the genomic basis of a particular disorder. Genetic analysis after 2000 did regularly take place at the level of DNA sequence analysis, but chromosome-level thinking and representation continued to be an important starting point for conversation and communication. As the use and power of molecular cytogenetic techniques expanded, reference back to the chromosomal infrastructure for diagnosis remained central to screening and analysis in medical genetics.

Microarray in the Clinic

In the years after the HGP, the resolution of array-CGH platforms used to screen the human genome for discrete deletions and duplications of chromosomal material increased rapidly. FISH analysis that had linked DNA sequence markers to specific locations along the human chromosomes provided large, well-characterized libraries of thousands of probes for use in array-CGH. Molecular cytogeneticists set up platforms to scan the entire human genome for deletions or duplications as small as 1 million base pairs, with the possibility of even higher-resolution analysis in specific regions. With more sensitive and broader-based microarray available, medical geneticists began applying this technique beyond the study of cancer cells to the clinical analysis of patients with various forms of intellectual disability of unknown origin.[36]

A few years into the 2000s, array-CGH was increasingly proving to be a powerful clinical tool for identifying the chromosomal addresses and genetic causes of various disorders. In one clinical microarray study, medical geneticists in the Netherlands identified either a deletion or a duplication in 5 of 20 patients affected by unidentified patterns of malformation involving intellectual disability. One deletion, on chromosome 7, was 8.6 million base pairs (Mb) long but could not be made visible under the microscope. This deletion overlapped the chromosomal address associated with Williams syndrome but was much larger than what was normally detected. The patient was also much more severely affected by intellectual disability and facial abnormalities than most with Williams syndrome, which these researchers suggested might imply that the deletion size and clinical severity were correlated. Abnormalities in other patients were closer to 2Mb in size and were found on various chromosomes. None of these findings were associated with a known disorder or potential candidate gene, and so medical geneticists could not verify the role of these apparent mutations in causing the patterns of malformation.[37]

In another study, published the next year, molecular cytogeneticists at Cambridge University used a similarly sensitive array-CGH platform to examine the genomes of 50 patients who showed patterns of malformation of unknown origin, involving intellectual disability. Twelve of these individuals were found to possess a chromosomal duplication or deletion, identified in various genomic locations. Two patients were found to possess deletions in the p36 region on chromosome 1, an area associated with a recently described clinical disorder known as 1p36 deletion syndrome. In 5 of the 12, researchers reported that the chromosomal abnormality had been inherited from a clinically normal parent, which they interpreted as meaning it was likely not involved in causing the pattern of malformation.[38]

Many medical geneticists believed that a significant number of the deletions and duplications that array-CGH analysis identified would prove to be benign, because they did not think the expression of most genes would be affected by one more or one fewer copy. The Cambridge researchers noted, "Currently, there is insufficient information available to determine how many of these copy number differences are of pathological significance. Additional data, gained both from the study of patients, and from normal individuals, will be needed to address this question." Unlike previous cytogenetic approaches, which could only detect relatively large chromosomal abnormalities that almost always had significant clinical impacts, many array-CGH findings represented relatively minor genetic variations, the pathology of which could not be interpreted until they were identified in multiple individuals. Medical geneticists recognized that much broader population studies were necessary for effective array-CGH interpretation.[39]

Beginning in the late 1990s, molecular cytogeneticist Lisa Shaffer became a leading figure in the clinical adoption and application of CGH microarray technology. At this time, Shaffer worked at the Baylor College of Medicine, where David Ledbetter was one of her colleagues in molecular human genetics. Her research included the use of FISH and microarray techniques to characterize 1p36 deletion syndrome. In 2003 Shaffer moved from Baylor to a position at Washington State University, where she also founded the biotechnology company Signature Genomic Laboratories along with physician Bassem Bejjani. After her move to Washington, Shaffer worked to bring broad-based array-CGH analysis into the clinical setting.

In a 2004 paper on recent advances in molecular cytogenetics, Shaffer noted, "CGH-based array will (i) increase the resolution . . . (ii) eliminate the need for targeted FISH experiments (i.e. needing to know where to FISH), and (iii) substantially decrease the labor involved." While FISH had improved cytogenetic resolution, it was a targeted approach that tested for only one mutation at a time, a feature that limited diagnostic capability. Array-CGH promised to circumvent these problems, but in the process it was also likely to produce unanticipated and uninterruptible results. As Shaffer put it, "The whole-genome approach that researchers are using is likely to generate data that may be difficult to interpret. Alterations in regions of the genome that do not have established clinical relevance will also be burdensome on the clinical cytogeneticists for useful interpretation."[40] Being able to conduct a molecular genetic test without having to know what they were looking for also meant that medical geneticists were not always going to know how to make sense of the results that they generated.

Mindful of these concerns, Signature Genomic Laboratories initially developed a targeted array-CGH test for molecular cytogenetics analysis, called the SignatureChip. The platform could be used to identify deletions and duplications of genetic material at 126 genomic locations that had previously been established as associated with intellectual disability or developmental delay. With this limited approach, Shaffer, Bejjani, and colleagues were pushing back against the moves of other molecular cytogeneticists toward fitting as many targets as possible into an array-CGH platform, a number that had already gone as high as 32,000 in one study.[41] In 2006 researchers at Signature Genomic Laboratories reported in the *Journal of Pediatrics* on a study involving 1,500 samples that had been sent to them from patients with malformation patterns, in which microscopic chromosomal analysis had identified no abnormalities. Subsequent testing with their targeted SignatureChip microarray identified 84 clinically relevant abnormalities and 14 more of unknown implication. While the yield of new results was relatively low (5.6%), the study represented an incremental improvement on traditional cytogenetic analysis, while primarily producing clinically interpretable results.[42] The SignatureChip was not likely to generate many new findings, but for medical geneticists it was a means of testing for many chromosomal abnormalities that were already known to cause genetic disorders, in a cheaper and more efficient manner than conducting individual FISH assays.

Microarray in the Prenatal Setting

As with other cytogenetic assays examined in these chapters, interest in the potential clinical application of array-CGH quickly turned to prenatal diagnosis and prevention. Shaffer and Bejjani were at the forefront of this move to prenatal diagnosis, along with their former colleagues at Baylor, led by physician Arthur Beaudet. FISH had already been widely adopted in the prenatal context to identify common trisomies and to test for microdeletions when a fetus was expected to be at risk for a specific disorder, such as familial DiGeorge syndrome. The use of FISH prenatally, however, was limited to these targeted applications. Molecular cytogeneticists presented array-CGH as promising to greatly expand the reach of prenatal diagnosis. However, existing concerns about the generation of abnormal findings of unknown clinical impact were only amplified in the prenatal context, when the observation of patterns of malformations in the fetus were limited to what could be made visible using ultrasound. In addition, parents were faced with the time-sensitive and difficult decision of pursuing prevention through abortion.[43]

Existing knowledge about the prevalence of copy number variants in clinically normal individuals raised the specter of frightened parents choosing to terminate healthy pregnancies based on the identification of potentially abnormal chromosomal findings. In a 2007 paper, bioethicist Evelyne Shuster referred to the adoption of such testing as a potential "roadblock for life." Shuster noted that historically, decisions about prenatal diagnosis were made one disorder at a time, usually beginning with Down syndrome, then adding others based on the assessment of risk. Prenatal microarray, she argued, threatened to disrupt this approach by making it both common and more cost effective to screen for hundreds or thousands of genetic abnormalities at once. Reflecting on the potential implications of this, Shuster warned, "Use of [prenatal] microarray genetic screening is likely to produce a flood of information that is overwhelming, anxiety-producing, inconclusive, and misleading. Genetic screening of adults can allow time to rescreen, to confirm findings, and to interpret or ignore results. But microarray genetic screening might allow only weeks or even days for a decision on whether to not to terminate a pregnancy." Microscopic chromosomal analysis in the prenatal setting had always involved the chance of uninterpretable results. The vastly increased resolution of array-CGH, however, was likely to greatly increase the incidence of uncertainty, possibly to the point where every fetus was found to have some potential genetic abnormality. Shuster suggested that parents, swayed by the fear of disease and the influence of genetic determinism, would choose to abort many more pregnancies, and as a result, "the quest for a healthy baby could cause parents to have no baby at all."[44]

Though stated in an extreme manner by Shuster, concerns about findings of unknown clinical significance certainly resonated with the molecular cytogeneticists at Signature Genomic Laboratories and Baylor. Both initially focused on providing relatively targeted microarrays for prenatal screening. The array-CGH test offered by Beaudet's laboratory tested for 100 clinically relevant chromosomal abnormalities at a cost of $2,000. While this was a high price, it was much cheaper and more efficient than conducting 100 separate FISH assays. Beaudet commented in a 2005 interview, "Our system allows you to do every known FISH test in the world at once."[45] While this microarray offered unprecedented genomic coverage, it was true that, as Shuster lamented, there was no choice available for parents to test prenatally for some disorders and not others.

The next year, Beaudet's laboratory reported on 98 pregnancies in which array-CGH had been performed using their platform alongside traditional cytogenetic analysis. Both tests identified four cases of trisomy 21 (Down syndrome). In 30 pregnancies, copy number variants were detected that were interpreted as normal, based on clinical experience and existing knowledge in the chromosomal infrastructure for diagnosis. Twelve other findings required follow-up analysis, as their clinical implications were uncertain. In nine of these cases, the same deletion or duplication was detected in one of the clinically normal parents, which Beaudet interpreted as meaning that it was benign. The remaining three findings of uncertain significance were novel to the fetus but did not appear to have any notable clinical impact. In these instances, parents were left with little information about the finding's significance.[46]

Signature Genomic Laboratories also began to offer targeted prenatal array-CGH in 2004. They published a report on their experience in 151 cases four years later. Most of these cases were referred for array-CGH analysis following an abnormal ultrasound result, which suggested a potential chromosome abnormality. Others were sent for testing by medical geneticists because of concerns stemming from family history or "patient anxiety." Shaffer's laboratory identified two clinically significant abnormalities, both involving chromosomal rearrangements that resulted in the gain or loss of genetic material. Twelve additional copy number variants were reported, and were interpreted as benign for reasons that were not explained. Signature Genomic Laboratories described just one of the genetic variants they had identified, among the 151 prenatal samples, as being of unknown clinical significance.[47]

By the time of their 2008 report, Shaffer and Bejjani's group had begun to move into prenatal testing with high-resolution, whole genome microarrays. One of their platforms screened for more than 1,500 loci, which were targeted to particular genomic regions of known clinical significance. Another testing option provided more extensive whole genome "backbone" coverage, comprising more than 100,000 probes spaced at 10kb intervals in certain genomic areas and 35kb everywhere else. Among 182 prenatal samples reported in 2012, the testing identified 7 clinically relevant genetic abnormalities and 17 copy number variants that were initially of unknown clinical significance.[48] All but one of these variants were found to occur also in one of the parents, and thus were regarded by Signature Genomic Laboratories as likely benign. As Shaffer's group later acknowledged, however, "inheritance from a parent often does not help assign the clinical relevance of the alteration because of the possibility of incomplete penetrance or variable expressivity. This is especially true in the laboratory setting in which clinical information on parents is limited."[49] Based on this, Signature Genomic Laboratories was more hesitant than other groups to immediately discount the potential significance of inherited copy number variants. Indeed, as in the case of DiGeorge syndrome, parents with the 22q11 deletion were sometimes undiagnosed, but their child with the same mutation was severely affected.

In a 2012 issue of *Prenatal Diagnosis,* Signature Genomic Laboratories and the Baylor group both published reports on their experience with more than 1,000 prenatal microarrays. Researchers from each location also contributed to an editorial outlining their results, which noted that the two groups reported quite different percentages of variants of unknown clinical significance: 4.2% for Signature and 1.6% for Baylor. As they acknowledged, the source of this difference came from what was counted as a genetic copy number variant of unknown significance. Signature Genomic Laboratories had included variants that they found to be inherited from a parent in the "unknown" category, whereas the Baylor team considered them benign. The Baylor group in fact had found variants of unknown significance in 4.2% of cases before additional parental studies were performed. How these findings were classified, and how parents were counseled about them, reflected local assumptions about the potential for intergenerational variability in clinical expression and likely influenced the decisions parents made.[50]

A larger multicenter study of prenatal microarray was also published in the *New England Journal of Medicine* in late 2012. It included findings from Baylor and Signature as well as many other study locations, such as Columbia, Drexel, and Northwestern universities. The study demonstrated the improved sensitivity of microarray over traditional cytogenetic analysis for prenatal studies but acknowledged that this came at the cost of increased counseling challenges and anxiety. Out of 3,822 samples, the researchers initially characterized 3.4% (130) as having uncertain findings. A team of experts from various centers discussed many of these uncertain variants in phone conferences and were able to characterize some as benign or likely pathogenic. Ultimately, however, in 56 cases (1.5%) the implications of the variant result were unknown during pregnancy. As the authors noted, "The interpretation of uncertain results will continue to require a close working relationship among laboratory directors, clinical geneticists, counselors, and practitioners."[51] While such interactions were possible within this well-funded, landmark study of prenatal microarray, the question of how this level of expert analysis would continue to be made available as the technology became widespread was not directly addressed. Indeed, as the clinical application of microarray in the most recent years highlighted, the scope and value of genomic screening was limited much less by technical factors or the cost of conducting testing than by the challenges of providing complete and accurate interpretation of the results.

Risks and Rewards of the Unknown

Alongside the multicenter prenatal microarray trial, social science researchers Marion Reiff and Barbara Bernhardt at the University of Pennsylvania conducted a study on how parents responded to and made sense of prenatal microarray results. Women reported being surprised and anxious after receiving positive results, and they worked quickly to try to make sense of them, given the short time frame for decision making about abortion. They often went online for information but had trouble interpreting genetic explanations or were misled by what they read. Even women who received the well-established diagnosis of a 22q11 deletion were faced with uncertainly about the disorder's severity, which could not be determined during pregnancy.

While women were counseled in advance that uncertain or complicated results could be generated by the microarray, they could not be adequately prepared for the reality of a variant or abnormal finding. Even following one or more counseling sessions after the positive result, many women were left with little certain information and little time to make a decision. As one respondent put it, "She's missing part of her seventh chromosome. However, they had absolutely no information as to what this meant." Although some received relatively clear diagnoses that would have been missed by traditional cytogenetic analysis, others were "blindsided" by uncertainty.[52] Either way, they faced a difficult and time-sensitive decision about whether to choose abortion.

No one was more aware of the challenging counseling sessions that some women who underwent prenatal microarray would face than the clinicians who offered and promoted the testing. Given its increased resolution, however, many providers regarded microarray to be a preferable prenatal option in comparison to traditional cytogenetic analysis. The Baylor group explained that "a certain number of newly discovered CNVs [copy number variants] have to be followed before valid conclusions about their overall clinical impact can be made." Looking at the big picture, they argued in 2012 that the long-term benefits were worth occasional anxiety, suggesting, "Over time, a small number of findings that were initially reported to have uncertain clinical significance can also become clearer as our clinical experience with chromosomal microarray analysis expands."[53] They promoted the continued contribution of variant findings to online resources such as the Database of Genomic Variants, so that novel findings would become increasingly rare and variants would be easier to interpret.

The application of microarray was not limited to the US context in the 2010s but was being conducted and grappled with by physicians and geneticists around the world. A significant global point of debate concerned what results should be reported or withheld from parents for their prenatal decision making. Some had argued that respect for patient autonomy required that all variant results, whether or not they could be reliably interpreted, should be reported to parents. Australian medical geneticist George McGillivray and colleagues argued, "Informing the woman that we are uncertain . . . and that we have no information to confirm or rule out pathogenicity, may cause her distress, but may also be relevant to her decision-making process. . . . Directive counseling to 'inform' the woman that termination is not appropriate, or to encourage her to continue the pregnancy may seem attractive here, but we argue that it is not ethically justified."[54] Others argued that a medical geneticist's role in prenatal counseling, as for other forms of diagnostic testing, was to determine which results were clinically relevant enough to be reported and what findings should be withheld. Molecular cytogeneticist David Ledbetter expressed such a view in 2012:

> I think managing the uncertainty [of prenatal microarray] is not as difficult as everybody thinks it is. The problem is that my laboratory colleagues feel compelled to report every trivial observation, when I think their responsibility is to do a clinical evaluation based on our knowledge today, and only report what's clinically interpretable and useful today. . . . It's sort of like a 'Goldilocks model' for interpretation of lab results. You don't want too much information because it raises anxiety, confusion, et cetera. And you don't want to miss anything. So, you want just right genetic information. And, it's the laboratory experts who should control how much data gets reported.[55]

From this perspective, the sensitivity of prenatal microarray and the likelihood that it would generate results of unknown clinical significance was less of a concern for prenatal diagnostic providers because they took on a more active role in deciding how the results were interpreted and presented to parents.

Some physicians and geneticists specifically pointed to the increased chance that high-resolution prenatal microarray would generate novel findings as a reason for encouraging its wider uptake. As one team of Dutch obstetricians and clinical geneticists put it in 2014, "To allow the discovery of new pathogenic CNVs, whole genome array platforms should be recommended in the prenatal setting."[56] Others pushed back against this viewpoint and questioned whether prenatal arrays that scan the whole genome, rather than target well-established mutations, should be presented to parents as a part of prenatal care. Another Dutch group argued in the same year that "undirected microarrays are a research tool rather than an instrument of clinical care. . . . Professionals who want to offer undirected arrays to their patients should make clear that the main purpose of doing so is research."[57] In 2014, the global medical genetics and obstetrics community continued to debate the question of who benefited from prenatal microarray, and how. Were those who underwent this testing being offered the best clinical option? Were they instead primarily subjects of research that was more likely to provide valuable information in 10 years than in 10 days? Opinions about how to address these concerns varied among physicians from the same country, as well as around the world.[58]

In the US context, the regulation of prenatal microarray was largely left to professional organizations. The American College of Obstetricians and Gynecologists (ACOG) weighed in on prenatal microarray with a new practice bulletin in December 2013. The policymakers recognized the potential improvements that microarray offered over traditional cytogenetic analysis, while acknowledging the anxiety associated with findings of unknown clinical significance. The group recommended cautious use of this test, suggesting that it was appropriate primarily for pregnancies in which a fetal structural abnormality had been identified under ultrasound. They left open the option of using prenatal microarray in all pregnancies but emphasized the need for extensive pre- and posttest counseling. Looking to the technique's future use, the practice bulletin noted, "Interpretation of results is expected to improve as knowledge of the human genome grows and the use of databases to link clinical findings and copy number variants becomes more robust."[59] In line with the growing number of active supporters and providers of prenatal microarray, ACOG expressed confidence that as the chromosomal infrastructure for diagnosis and prevention continued to grow and become more densely filled in, the value of prenatal microarray for the identification and prevention of genetic disorders would improve. With microarray technology, prenatal diagnosis had expanded beyond targeted disorders and was moving toward a new gold standard of DNA-level whole genome screening.

Conclusions

Human cytogenetic analysis, from its origins in the 1950s, had always been a form of whole genome analysis. Throughout the twentieth century, however, it was a very blunt instrument, which medical geneticists could use only to make visible large chromosomal abnormalities. In the 1980s high-resolution cytogenetic analysis offered limited improvements but was not likely to identify new mutations. Human cytogeneticists were never successful, as Jorge Yunis, David Ledbetter, and others had hoped, in screening for new disease-causing microdeletions under the microscope.[60] The introduction of microarray in the years after 2000 combined the density of molecular analysis with the breadth of traditional cytogenetic observation. With this platform, medical geneticists were able to conduct whole genome analysis at an extremely high resolution. While microarray represented a significant technical advance in genetic testing, in order to make sense of the results it generated, medical geneticists continued to rely on knowledge that had been compiled, in a piecemeal fashion since 1970, in the chromosomal infrastructure for diagnosis. Used together, microarray platforms and the content of the chromosomal infrastructure facilitated molecular-level whole genome analysis. Once molecular cytogeneticists, in academic laboratories, clinical settings, and private corporations, had harnessed the power of existing diagnostic technologies and data sets, they presented a product to physicians and parents that one company promised would finally provide "the whole picture" for prenatal diagnosis.[61]

Each of the studies described in this book, and thousands more that took place concurrently since the 1960s, contributed to the slow and steady development of the chromosomal infrastructure for diagnosis and prevention. In many instances, medical geneticists promised that their individual studies would bring improvements in the understanding and treatment of a clinical disorder through the identification of its chromosomal address and genomic cause. Whether or not identifying a disorder's place among the genome's morbid anatomy succeeded in bringing these promised results, it likely did contribute to prevention, adding to the potential outcomes and decisions to be made based on prenatal microarray.

During the first two decades of the 2000s, clinical studies continued to add new information to the chromosomal infrastructure for diagnosis and prevention, which by then was housed and accessed primarily through the genome browsers described in this chapter. In addition to targeted studies of particular disorders, the results of microarray analysis were added to the infrastructure, as medical geneticists interpreted newly identified variants as pathological or benign, based on long-term clinical follow-up in the postnatal context, if it occurred. While these findings could help them to clarify decisions to be made in future pregnancies, many newly recognized variants provided only uncertainty and anxiety for parents in the immediate prenatal context.

There was an inherent risk and a potential reward in attempts to "get the whole picture." Careful genetic counseling was meant to make parents aware of these potential problems so that they could make an informed decision about what testing to pursue. However, it was nearly impossible to prepare or warn parents for the degree of potential unknowns that could result from prenatal microarray.[62] Parents might very well be left either anxious or comforted by the results. Either way, these parents were part of a larger experiment, whether they realized it or not, the long-term aims of which were unclear even to those who were conducting it. Just as researchers in the 1970s could not have fully appreciated the extent to which their studies of Prader-Willi or DiGeorge syndrome would contribute to the development of the infrastructure for diagnosis, parents and physicians in the 2010s could not grasp how the results from prenatal microarray would feed into the preventive measures, and perhaps treatments, of the future.

The marketplace for prenatal diagnosis after 2010 was awash with companies and medical geneticists who were promising to provide "the whole picture" through genome-wide molecular screening. What they were selling was much more than a simple test: it was a means to many ends, some of which were yet to be defined. Once collected, cellular and genetic samples often did not disappear or become outdated. Instead, they were transformed in a variety of ways, into useful cell lines, the components of genomic libraries, and massive databases of genetic variants. These data and resources were regularly funneled into basic research and clinical analysis that were not even imagined at the time of collection. Parents may have been offered the chance to consent to having their sample included in research, but what were they even consenting to? There was no way for anyone to know. Whether or not they received the results for which they were looking, parents could be sure, even if they were not thoroughly informed of it, that any notable results from genetic analysis were likely to be retained and to become part of research and prevention regimes for decades to come.

Epilogue

The Genomic Gaze

As medical geneticists worked to increase the resolving power of genomic analysis during the closing decades of the twentieth century, to identify many new disease-causing mutations at the molecular level, historians and sociologists began to debate a provocative suggestion made in 1990 by scholar Donna Haraway. She proposed, "It is time to write *The Death of the Clinic*."[1] Had recent advances in the molecular diagnosis and understanding of disease, based in the laboratory, made clinical investigation redundant? Had the analysis of genomic texts replaced that of bodies? While many scholars resisted going so far as to proclaim the "death of the clinic," some pointed to an epistemic shift in medicine that had taken shape around the introduction of new molecular approaches since the 1980s. Sociologists Adele Clarke and Nikolas Rose, among others, argued that practitioners of late-twentieth-century biomedicine visualized life at the molecular level, a view that was replacing the clinical gaze of medicine with a new style of thought and way of seeing, which they termed the "molecular gaze." As Rose put it in 2007, the clinical gaze of medicine "has been supplemented, if not supplanted" by the molecular gaze of biomedicine.[2]

Other scholars contested the suggestion that medicine had experienced an epistemic break during the late postwar period. Sociologists Katie Featherstone and Paul Atkinson argued instead, "The clinic seems permanently poised between tradition and innovation, between the *longue durée* of its core characteristics, and rapidly changing scientific knowledge." Thus, claims of "wholesale transformations . . . seem premature at best."[3] These scholars highlighted the resilience of earlier modes of thought and practice in medicine. In the debate over whether the molecular gaze had supplemented or supplanted the clinical gaze, Featherstone and Atkinson came down on the side of supplementation. Going further, sociologist Joanna Latimer, drawing on ethnography in dysmorphology clinics, claimed that the rise of molecular genetics brought about a "rebirth" of the clinic as a key site of knowledge production in genetics.[4]

Sociologists Vololona Rabeharisoa and Pascale Bourret supported a similar position, suggesting, based on their own clinical ethnographic analysis, that "the presence or absence of mutations cannot be considered as *the* 'objective proof' of the existence or absence of a particular syndrome." They noted that mutations were themselves clinical objects that influenced medical interpretation but did not provide definitive diagnostic answers on their own. The identification of disorders, they argued, did not take place exclusively in the laboratory or clinic, but rather in "bioclinical collectives" that included diverse expertise. Historians Ilana Löwy and Soraya de Chadarevian also demonstrated, through extensive archival and ethnographic research, that postwar biomedical research involved a two-way street of knowledge exchange between clinicians and basic biologists.[5] Adding to this narrative, historian Peter Keating and sociologist Alberto Cambrosio traced the sharing of research materials and standards among the laboratory and the clinic, while highlighting the often-overlooked role of clinical questions and concerns in shaping basic biomedical research aims and outcomes.[6]

Throughout this book, I trace how the clinical gaze was influenced by the one mutation–one disorder ideal of postwar medical genetics. The practices of dysmorphology, since its origins around 1970, were shaped by the assumption that discrete patterns of malformation could be discerned that had a single, often genetic, basis (chapter 1). During the same decade, medical geneticists began to map these distinct clinically described genetic disorders within the human genome, making use of an infrastructure that was built upon the nomenclature of banded human chromosomes (chapter 2). By the early 1980s, medical geneticists had associated many patterns of malformation with specific chromosomal addresses in the human genome's morbid anatomy. In this way, the clinical gaze was brought to human chromosomes and the human genome, which were presented in the publications and presentations of influential medical geneticists such as Victor McKusick and Charles Scriver as the domain of physicians and geneticists alike. Physicians increasingly understood many clinical disorders as having their own specific and knowable address within the chromosomal infrastructure of the human genome (chapter 3).

During the 1980s and 1990s, human cytogeneticists, in collaboration with molecular biologists, worked to greatly increase the resolution and ease of chromosomal analysis and, with this, the density of genomic knowledge (chapter 4). It was during this era, Rose and Clarke later argued, that the molecular gaze emerged and began to reshape how disorders were diagnosed and prevented. In situ hybridization techniques such as FISH seemed to finally bring greater clarity to clinical diagnosis. They further supported the one mutation–one disorder assumptions of medical geneticists but also led to new social conflicts over questions of expertise, naming, and institutional status in genetic medicine (chapter 5). With the completion of the Human Genome Project in 2000, molecular cytogeneticists greatly expanded the reach of molecular-level analysis through the development of genome-wide microarray. While these testing platforms contributed to molecular-level diagnosis of disorders, medical geneticists continued to link the results back to chromosomal conceptions and maps of the genome, because this level of organization, analysis, and communication was more familiar and legible for them and their patients (chapter 6).

I argue that the era between the 1970s and the present did not reflect a transition from one style of thought to another—from the clinical gaze to the molecular gaze—but rather the ongoing development of a genomic gaze, which incorporated both. Beginning in the 1950s, medical geneticists looked to the chromosomes for a "glimpse" of the entire human genome.[7] After 2010, even when medical geneticists had the entire DNA sequence of the human genome at their fingertips, they continued to look back to the comprehensive vision of the genome embodied by banded chromosomal ideograms, first developed in the 1970s. The genomic gaze spanned many levels of perception and analysis, but its focus was continuously channeled back to chromosome-level conceptions and depictions. When medical geneticists considered the various minute features of a particular genomic region, they often envisioned a chromosomal ideogram. When they turned to a genome browser to examine a particular gene in that region, a banded chromosome was there to orient them within the genome. Similarly, when physicians worked to diagnose a particular clinical pattern of malformation, they sought also to locate it at a specific, standardized chromosomal address within the genome's morbid anatomy.

The genomic gaze of medical genetics integrated knowledge from bodies in the clinic and molecules in the laboratory into a chromosomal infrastructure for diagnosis and prevention, to which data was continuously added between the 1970s and the 2010s. This infrastructure brought together the whole genome view facilitated by microscopic chromosome-level analysis and the highly specific and targeted examination of molecular mutations at the DNA level. Even with various advancements in molecular technology, the central focus of the genomic gaze continued to be at the level of banded chromosomal ideograms. This was the common ground, where medical geneticists from clinical, cytogenetic, and molecular subspecialties could think and work collaboratively with data and understandings from radically different sources. Though the density of available genomic data had expanded at an increasing rate, conceptions of the total genome remained focused on chromosome-level depictions. Chromosomes were the level at which all practitioners shared a tangible and legible grasp of the genome's expansive content.

In the post-2000 era of microarray, the genomic gaze incorporated the totalizing breadth of the clinical gaze and the vast depth of the molecular gaze into one style of thought. Medical geneticists sought to compile and explain each genetic variant and clinical pattern of malformation identified in every person. Under the genomic gaze, all conditions were at once clinical and molecular. The focal point of medical genetics analysis was the chromosomal addresses to which all of these diverse findings were linked. Even as medical geneticists acknowledged that the human genome functioned in complex and multifactorial ways, their long-standing one mutation–one disorder ideal continued to orient the genomic gaze. Multiple genes could be shown to shape a clinical outcome, but one chromosomal address remained the primary means of defining the cause of a disorder and facilitating its diagnosis. While medical geneticists had compiled the chromosomal infrastructure little by little, beginning in the 1970s, through a focus on improving the understanding, diagnosis, and treatment of individual genetic disorders, five decades later their infrastructure as a whole had become an edifice for prevention.

Notes

Introduction. Pursuing a Better Birth

1. On the issue of prenatal choice and pressure to test in the United Kingdom, see Claire Williams, Priscilla Alderson, and Bobbie Farsides, "Too Many Choices? Hospital and Community Staff Reflect on the Future of Prenatal Screening," *Social Science and Medicine* 55, no. 5 (2002): 743–53.

2. For more on infrastructures in science and medicine, see Paul N. Edwards, "Infrastructure and Modernity: Force, Time, and Social Organization in the History of Sociotechnical Systems," in *Modernity and Technology*, ed. Thomas J. Misa, Philip Brey, and Andrew Feenberg (Cambridge, MA: MIT Press, 2004), 185–226; Geoffrey C. Bowker and Susan Leigh Star, *Sorting Things Out: Classification and Its Consequences* (Cambridge, MA: MIT Press, 1999); Susan Leigh Star and Karen Ruhleder, "Steps toward an Ecology of Infrastructure: Design and Access for Large Information Spaces," *Information Systems Research* 7, no. 1 (1996): 111–34; Angela N. H. Creager and Hannah Landecker, "Technical Matters: Method, Knowledge and Infrastructure in Twentieth-Century Life Science," *Nature Methods* 6, no. 10 (2009): 701–5.

3. Nathaniel Comfort, *The Science of Human Perfection: Heredity and Health in American Biomedicine* (New Haven, CT: Yale University Press, 2012); Diane B. Paul, *Controlling Human Heredity: 1865 to the Present* (Atlantic Highlands, NJ: Humanities Press, 1995); Daniel J. Kevles, *In the Name of Eugenics: Genetics and the Uses of Human Heredity* (New York: Alfred A. Knopf, 1985); Alexandra Minna Stern, *Eugenic Nation: Faults and Frontiers of Better Breeding in Modern America* (Berkeley: University of California Press, 2005); Kenneth M. Ludmerer, *Genetics and American Society: A Historical Appraisal* (Baltimore: Johns Hopkins University Press, 1972); Paul A. Lombardo, *Three Generations of Imbeciles: Eugenics, the Supreme Court, and Buck V. Bell* (Baltimore: Johns Hopkins University Press, 2008); Martin S. Pernick, *The Black Stork: Eugenics and the Death of "Defective" Babies in American Medicine and Motion Pictures since 1915* (New York: Oxford University Press, 1996).

4. Nathaniel Comfort, "'Polyhybrid Heterogeneous Bastards': Promoting Medical Genetics in America in the 1930s and 1940s," *Journal of the History of Medicine and Allied Sciences* 61, no. 4 (2006): 415–55; Ruth Schwartz Cowan, *Heredity and Hope: The Case for Genetic Screening* (Cambridge, MA: Harvard University Press, 2008); Yoshio Nukaga and Alberto Cambrosio, "Medical Pedigrees and the Visual Production of Family Disease in Canadian and Japanese Genetic Counselling Practice," *Sociology of Health & Illness* 19, no. 19B (1997): 29–55; Reed E. Pyeritz, "The Family History: The First Genetic Test, and Still Useful after All Those Years?," *Genetics in Medicine* 14, no. 1 (2011): 3–9.

5. Stern, *Eugenic Nation;* Comfort, *Science of Human Perfection;* Peter S. Harper, *A Short History of Medical Genetics* (Oxford: Oxford University Press, 2008); Diane B. Paul,

The Politics of Heredity: Essays on Eugenics, Biomedicine, and the Nature-Nurture Debate (Albany: State University of New York Press, 1998); Alexandra Minna Stern, *Telling Genes: The Story of Genetic Counseling in America* (Baltimore: Johns Hopkins University Press, 2012).

6. Troy Duster, *Backdoor to Eugenics* (New York: Routledge, 1990); Marc Lappe, "How Much Do We Want to Know about the Unborn?," *Hastings Center Report* 3, no. 1 (1973): 8–9; Susan Lindee, *Moments of Truth in Genetic Medicine* (Baltimore: Johns Hopkins University Press, 2005); Abby Lippman, "Led (Astray) by Genetic Maps: The Cartography of the Human Genome and Health Care," *Social Science & Medicine* 35, no. 12 (1992): 1469–76; Rayna Rapp, *Testing Women, Testing the Fetus: The Social Impact of Amniocentesis in America* (New York: Routledge, 1999); Barbara Katz Rothman, *The Tentative Pregnancy: Prenatal Diagnosis and the Future of Motherhood* (New York: Viking, 1986).

7. Nicholas Agar, *Liberal Eugenics: In Defence of Human Enhancement* (New York: Blackwell, 2004); Stern, *Telling Genes;* Katherine Castles, "'Nice Average Americans': Postwar Parents' Groups and the Defense of the Normal Family," In *Mental Retardation in America: A Historical Reader,* ed. Steven Noll and James W. Trent (New York: New York University Press, 2004), 351–70; Erik Parens and Adrienne Asch, eds., *Prenatal Testing and Disability Rights* (Washington, DC: Georgetown University Press, 2000); David Wright, *Downs: The History of a Disability* (New York: Oxford University Press, 2011).

8. Comfort, *Science of Human Perfection;* V. A. McKusick, "The Growth and Development of Human Genetics as a Clinical Discipline," *American Journal of Human Genetics* 27, no. 3 (1975): 261–73.

9. Comfort, *Science of Human Perfection;* Harper, *Short History of Medical Genetics;* Kevles, *In the Name of Eugenics.*

10. Ilana Löwy, "How Genetics Came to the Unborn: 1960–2000," *Studies in History and Philosophy of Biological and Biomedical Sciences* 47 (2014): 290–99; D. W. Smith, "Dysmorphology (Teratology)," *Journal of Pediatrics* 69, no. 6 (1966): 1150–69, 1151; D. W. Smith, *Recognizable Patterns of Human Malformation* (Philadelphia: W. B. Saunders, 1970). A decade later, Smith published a second compendium on recognizable patterns of human "deformation" caused by environmental and mechanical factors in the womb. Smith's decision to develop this parallel reference text highlighted his desire to maintain the focus of dysmorphology on genetic disorders and to distinguish these from environmentally caused conditions such as fetal alcohol syndrome, which he helped to first identify and name in 1973. K. L. Jones and D. W. Smith, "Recognition of the Fetal Alcohol Syndrome in Early Infancy," *Lancet* 302, no. 7836 (1973): 999–1001; D. W. Smith, "Recognizable Patterns of Human Deformation: Identification and Management of Mechanical Effects on Morphogenesis," *Major Problems in Clinical Pediatrics* 21 (1981): 1–151.

11. R. L. Brent, "Biography of Josef Warkany," *Teratology* 25, no. 2 (1982): 137–51; F. C. Fraser, "Of Mice and Children: Reminiscences of a Teratogeneticist," *American Journal of Medical Genetics Part A* 146A, no. 17 (2008): 2179–2202; H. Kalter, "Josef Warkany 1902–1992," *Teratology* 48, no. 1 (1993): 1–3; E. Passarge, "Josef Warkany," *Lancet* 340, no. 8819 (1992): 602; F. C. Fraser, "The William Allan Memorial Award Address: Evolution of a Palatable Multifactorial Threshold Model," *American Journal of Human Genetics* 32, no. 6 (1980): 796–813.

12. James Crow, interview by Peter Harper, October 24, 2005, Interviews with Human and Medical Geneticists series, Special Collections and Archives, Cardiff University, Cardiff, UK; Harper, *Short History of Medical Genetics;* G. Neri and J. M. Opitz, "Down Syndrome: Comments and Reflections on the 50th Anniversary of Lejeune's Discovery," *American Journal of Medical Genetics* 149A, no. 12 (2009): 2647–54; K. Patau, D. W. Smith, E. Therman, S. L. Inhorn, and H. P. Wagner, "Multiple Congenital Anomaly Caused by an Extra Autosome," *Lancet* 1, no. 7128 (1960): 790–93.

13. J. M. Aase, "The Dysmorphology Detective," *Pediatric Annals* 10, no. 7 (1981): 38–43, 38; Harper, *Short History of Medical Genetics.*

14. Smith, "Dysmorphology," 1155; Aase, "Dysmorphology Detective"; Elaine Zackai, interview by author, Philadelphia, November 10, 2011; Michael Mennuti, interview by author, Philadelphia, November 7, 2011.

15. J. G. Hall, J. Allanson, K. Gripp, and A. Slavotinek, *Handbook of Physical Measurements,* 2nd ed. (New York: Oxford University Press, 2006); Sharrona Pearl, *About Faces: Physiognomy in Nineteenth-Century Britain* (Cambridge, MA: Harvard University Press, 2010); Lorraine Daston and Peter Galison, "The Image of Objectivity," *Representations* 40, no. 3 (1992): 81–128; Waltraud Ernst and Bernard Harris, eds., *Race, Science, and Medicine, 1700–1960* (London: Routledge, 1999).

16. Smith, "Dysmorphology," 1165.

17. Ibid.; Smith, *Recognizable Patterns of Human Malformation;* Löwy, "How Genetics Came."

18. Smith, "Dysmorphology"; D. Levenson, "David W. Smith Workshop Celebrates 30 Years of Discovery," *American Journal of Medical Genetics Sequence* (2010): vii–ix; Katie Featherstone, Joanna Latimer, Paul Atkinson, Daniella T. Pilz, and Angus Clarke, "Dysmorphology and the Spectacle of the Clinic," *Sociology of Health & Illness* 27, no. 5 (2005): 551–74; Michael Baraitser, interview by Peter Harper, March 1, 2005, Interviews with Human and Medical Geneticists series, Special Collections and Archives, Cardiff University, Cardiff, UK; Dian Donnai, interview with Peter Harper, February 6, 2007, ibid.

19. Joseph November, *Biomedical Computing: Digitizing Life in the United States* (Baltimore: Johns Hopkins University Press, 2012); Maria Jesus Santesmases, "The Human Autonomous Karyotype and the Origins of Prenatal Testing: Children, Pregnant Women and Early Down's Syndrome Cytogenetics, Madrid, 1962–1975," *Studies in the History and Philosophy of Biological and Biomedical Sciences* 47 (2014): 142–53; Mauro Turrini, "Continuous Grey Scales versus Sharp Contrasts: Styles of Representation in Italian Clinical Cytogenetics Laboratories," *Human Studies* 35, no. 1 (2012): 1–25; Rapp, *Testing Women, Testing the Fetus;* Löwy, "How Genetics Came."

20. Uta Francke, interview by author, Palo Alto, CA, February 27, 2012; Denver Study Group, "A Proposed Standard System of Nomenclature of Human Mitotic Chromosomes," *Lancet* 275, no. 7133 (1960): 1063–65; Paris Conference (1971), "Standardization in Human Cytogenetics," *Birth Defects Original Article Series* 8, no. 7 (1972): 1–46.

21. H. J. Muller, "Our Load of Mutations," *American Journal of Human Genetics* 2, no. 2 (1950): 111–76; Soraya de Chadarevian, "Mutations in the Nuclear Age," in *Making Mutations: Objects, Practices, Contexts, Preprint 393,* ed. Luis Campos and Alexander von Schwerin (Berlin: Max Plank Institute for the History of Science, 2010), 179–88, 180.

22. Kevles, *In the Name of Eugenics*, 241; Malcolm Jay Kottler, "From 48 to 46: Cytological Technique, Preconception, and the Counting of Human Chromosomes," *Bulletin of the History of Medicine* 48, no. 4 (1974): 465–502.

23. Aryn Martin, "Can't Any Body Count? Counting as an Epistemic Theme in the History of Human Chromosomes," *Social Studies of Science* 34, no. 6 (2004): 923–48; Kevles, *In the Name of Eugenics*.

24. J. Lejeune, M. Gautier, and R. Turpin, "Mongolism: A Chromosomal Disease (Trisomy)," *Bulletin de l'Academie Nationale de Medecine* 143, nos. 11–12 (1959): 256–65. Often overlooked was the contribution of Lejeune's coauthor, human cytogeneticist Marthe Gautier, who prepared and first examined the microscopic slides from a Mongolism patient's tissue sample. Gautier later argued that she had noted the extra chromosome but lacked the microscopic power and photographic equipment needed to report on it, and so she turned the slides over to Lejeune to produce these images. Soon thereafter, Lejeune reported the discovery as his own. Debates over how much credit Gautier deserved for the discovery continued past 2010, two decades after Lejeune's death. Marthe Gautier and Peter S. Harper, "Fiftieth Anniversary of Trisomy 21: Returning to a Discovery," *Human Genetics* 126 (2009): 317–24; Elisabeth Pain, "After More than 50 Years, a Dispute over Down Syndrome Discovery," *Science* 343 (2014): 720–21.

25. Jean-Paul Gaudillière, "Bettering Babies: Down's Syndrome, Heredity and Public Health in Post-War France and Britain," in *Images of Disease: Science, Public Policy, and Health in Post-War Europe*, ed. Ilana Löwy and John Krige (Luxemburg: Office for Official Publications of the European Communities, 2001), 89–108; Maria Jesus Santesmases, "Size and the Centromere: Translocations and Visual Cultures in Early Human Genetics," in Campos and von Schwerin, *Making Mutations*, 189–208.

26. V. A. McKusick, "A History of Medical Genetics," in *Emery and Rimoin's Principles and Practice of Medical Genetics* (New York: Elsevier Health Sciences, 1997), 1–30; Harper, *Short History of Medical Genetics*.

27. Smith, "Dysmorphology," 1159; Kevles, *In the Name of Eugenics;* Kurt Hirschhorn, interview by author, New York, January 26, 2012; Wright, *Downs*.

28. Cowan, *Heredity and Hope*.

29. Ilana Löwy, "Prenatal Diagnosis: The Irresistible Rise of the 'Visible Fetus,'" *Studies in the History and Philosophy of Biology and Biomedical Sciences* 47 (2014): 290–99, 290; Löwy, "How Genetics Came"; Diane B. Paul and Jeffrey P. Brosco, *The PKU Paradox: A Short History of a Genetic Disease* (Baltimore: Johns Hopkins University Press, 2013).

30. R. G. Resta, "Historical Aspects of Genetic Counseling: Why Was Maternal Age 35 Chosen as the Cut-Off for Offering Amniocentesis?," *Medicina Nei Secoli Art E Scienza* 14, no. 3 (2002): 793–811; Stern, *Telling Genes*.

31. American College of Obstetricians and Gynecologists, "ACOG Practice Bulletin No. 88, December 2007: Invasive Prenatal Testing for Aneuploidy," *Obstetrics and Gynecology* 110, no. 6 (2007): 1459–67; American College of Obstetricians and Gynecologists, "ACOG Practice Bulletin No. 77: Screening for Fetal Chromosomal Abnormalities," *Obstetrics and Gynecology* 109, no. 1 (2007): 217–27; Löwy, "Prenatal Diagnosis," 290.

32. McKusick, "History of Medical Genetics," 8; de Chadarevian, "Mutations in the Nuclear Age," 180.

33. Lindee, *Moments of Truth;* Comfort, *Science of Human Perfection.*

34. V. A. McKusick, *Mendelian Inheritance in Man: Catalogs of Autosomal Dominant, Recessive, and X-Linked Phenotypes* (Baltimore: Johns Hopkins Press, 1966).

35. Ibid., x.

36. Ibid., ix.

37. Ibid.

38. V. A. McKusick, "On Lumpers and Splitters, or the Nosology of Genetic Disease," *Birth Defects* 5, no. 1 (1969): 23–32. Similar instances of etiological lumping occurred in bacteriology, for instance when microbiologists argued that pinta, yaws, and syphilis were all caused by one microorganism, *Treponema pallidum.* C. J. Hackett, "On the Origin of the Human Treponematoses (Pinta, Yaws, Endemic Syphilis and Venereal Syphilis)," *Bulletin of the World Health Organization* 29 (1963): 7–41.

39. Like other productive experimental systems, human genetic disease research drew many new participants during the 1980s and 1990s because it regularly produced new epistemic entities. Hans-Jorg Rheinberger, *Toward a History of Epistemic Things: Synthesizing Proteins in the Test Tube* (Stanford, CA: Stanford University Press, 1997).

40. R. E. Pyeritz, "Marfan Syndrome: 30 Years of Research Equals 30 Years of Additional Life Expectancy," *Heart* 95, no. 3 (2009): 173–75.

41. Charles E. Rosenberg, "Framing Disease: Illness, Society and History," in *Framing Disease: Studies in Cultural History,* ed. Charles E. Rosenberg and Janet Golden (New Brunswick, NJ: Rutgers University Press, 1992), xiii–xvi, xiii.

42. Robert A. Aronowitz, *Making Sense of Illness: Science, Society, and Disease* (Cambridge: Cambridge University Press, 1998); Georges Canguilhem, *The Normal and the Pathological* (New York: Zone, 1991); Michel Foucault, *Birth of the Clinic: An Archaeology of Medical Perception* (New York: Pantheon, 1973).

43. Andrew J. Hogan, "Medical Eponyms: Patient Advocates, Professional Interests, and the Persistence of Honorary Naming," *Social History of Medicine,* forthcoming in 2016; Wright, *Downs;* Parens and Asch, *Prenatal Testing and Disability Rights;* Allison C. Carey, *On the Margins of Citizenship: Intellectual Disability and Civil Rights in Twentieth-Century America* (Philadelphia: Temple University Press, 2010).

44. V. A. McKusick, *Mendelian Inheritance in Man: Catalogs of Autosomal Dominant, Recessive, and X-Linked Phenotypes,* 5th ed. (Baltimore: Johns Hopkins University Press, 1978), xx.

45. Hirschhorn interview; J. M. Opitz, "Associations and Syndromes: Terminology in Clinical Genetics and Birth Defects Epidemiology: Comments on Khoury, Moore, and Evans," *American Journal of Medical Genetics* 49, no. 1 (1994): 14–20.

46. After 2000 there were rare instances of syndrome delineation beginning with a common mutation. Daniel Navon, "Genomic Designation: How Genetics Can Delineate New, Phenotypically Diffuse Medical Categories," *Social Studies of Science* 20, no. 10 (2011): 1–24.

47. Daniel J. Kevles and Leroy E. Hood, eds., *Code of Codes: Scientific and Social Issues in the Human Genome Project* (Cambridge, MA: Harvard University Press, 1992); Dorothy Nelkin and M. Susan Lindee, *The DNA Mystique: The Gene as a Cultural Icon* (New York: W. H.

Freeman, 1995); Lily E. Kay, *Who Wrote the Book of Life?: A History of the Genetic Code* (Stanford, CA: Stanford University Press, 2000).

48. Lippman, "Led (Astray) by Genetic Maps."

49. Dorothy Nelkin and Laurence Tancredi, *Dangerous Diagnostics: The Social Power of Biological Information* (New York: Basic Books, 1989); Parens and Asch, *Prenatal Testing and Disability Rights;* Evelyne Shuster, "Microarray Genetic Screening: A Prenatal Roadblock for Life?," *Lancet* 369, no. 9560 (2007): 526–29; Keith Wailoo and Stephen Pemberton, *The Troubled Dream of Genetic Medicine: Ethnicity and Innovation in Tay-Sachs, Cystic Fibrosis, and Sickle Cell Disease* (Baltimore: Johns Hopkins University Press, 2006); Williams, Alderson, and Farsides, "Too Many Choices?"; Rothman, *Tentative Pregnancy;* Rapp, *Testing Women, Testing the Fetus;* Duster, *Backdoor to Eugenics.*

50. D. D. Weaver, *Catalog of Prenatally Diagnosed Conditions,* 2nd ed. (Baltimore: Johns Hopkins University Press, 1992).

51. Parens and Asch, *Prenatal Testing and Disability Rights,* 23–29.

52. Parens and Asch, *Prenatal Testing and Disability Rights;* Carey, *On the Margins of Citizenship;* Noll and Trent, *Mental Retardation in America;* Wright, *Downs.*

53. Parens and Asch, *Prenatal Testing and Disability Rights;* Shuster, "Microarray Genetic Screening," 528.

Chapter 1. Genetics Detectives

1. L. S. Penrose, *A Clinical and Genetic Study of 1280 Cases of Mental Defect (The Colchester Survey)* (London: Medical Research Council, 1938); Daniel J. Kevles, *In the Name of Eugenics: Genetics and the Uses of Human Heredity* (New York: Alfred A. Knopf, 1985). The medical terms used to describe intellectual impairment were evolving throughout the nineteenth and twentieth centuries. They included *idiocy, feeblemindedness, mental retardation,* and, beginning around the turn of the twenty-first century, *intellectual disability.* In this book, I use the terminology of my actors within a specific time and place. Robert L. Schalock, "The Renaming of Mental Retardation: Understanding the Change to the Term Intellectual Disability," *Intellectual and Developmental Disabilities* 45, no. 2 (2007): 116–24.

2. L. S. Penrose, *The Biology of Mental Defect* (New York: Grune and Stratton, 1949).

3. J. B. S. Haldane, preface to Penrose, *Biology of Mental Defect,* v–vii.

4. Ellen Dwyer, "The State and the Multiply Disadvantaged: The Case of Epilepsy," in *Mental Retardation in America: A Historical Reader,* ed. Steven Noll and James Trent (New York: New York University Press, 2004), 258–80, 265–73.

5. V. A. McKusick, "A History of Medical Genetics," in *Emery and Rimoin's Principles and Practice of Medical Genetics* (New York: Elsevier Health Sciences, 1997), 1–30, 8; Ilana Löwy, "How Genetics Came to the Unborn: 1960–2000," *Studies in History and Philosophy of Biological and Biomedical Sciences* 47 (2014): 290–99. As Löwy insightfully points out, chromosomal disorders such as Down syndrome were not genetic in the traditional sense of being inherited, but they were still co-opted for prenatal testing by medical geneticists and were central to the field's significant growth in the late twentieth century.

6. H. A. Lubs and J. H. Salmon, "The Chromosomal Complement of Human Solid Tumors: II, Karyotypes of Glial Tumors," *Journal of Neurosurgery* 22, no. 2 (1965):

160–68; W. M. Court Brown, K. E. Buckton, P. A. Jacobs, I. M. Tough, E. V. Kuenssberg, and J. D. E. Knox, *Chromosome Studies on Adults* (London: Cambridge University Press, 1966).

7. H. A. Lubs, "A Marker X Chromosome," *American Journal of Human Genetics* 21, no. 3 (1969): 231–44, 233; H. A. Lubs and F. H. Ruddle, "Chromosomal Abnormalities in the Human Population: Estimation of Rates Based on New Haven Newborn Study," *Science* 169, no. 3944 (1970): 495–97.

8. Lubs, "Marker X Chromosome," 234.

9. Yoshio Nukaga and Alberto Cambrosio, "Medical Pedigrees and the Visual Production of Family Disease in Canadian and Japanese Genetic Counselling Practice," *Sociology of Health & Illness* 19, no. 19B (1997): 29–55.

10. Lubs, "A Marker X Chromosome."

11. Susan Lindee, *Moments of Truth in Genetic Medicine* (Baltimore: Johns Hopkins University Press, 2005).

12. Lubs, "Marker X Chromosome." Additional laboratory work included radiographic labeling, linkage studies, and precise arm length measurements of the abnormal chromosome.

13. L. S. Penrose, *Outline of Human Genetics,* 2nd ed. (New York: John Wiley, 1963), 136; V. A. McKusick, *Mendelian Inheritance in Man: Catalogs of Autosomal Dominant, Recessive, and X-Linked Phenotypes,* 2nd ed. (Baltimore: Johns Hopkins University Press, 1968), xi.

14. Lubs, "Marker X Chromosome," 231.

15. Ibid., 243.

16. M. W. Steele and W. R. Breg, "Chromosome Analysis of Human Amniotic-Fluid Cells," *Lancet* 287, no. 7434 (1966): 383–85. Chromosomal identification was limited to large gains and losses of genetic material, which often resulted in miscarriage before they were detected. Many biochemical tests for particular disorders were also in varying stages of development at this time. J. E. Brody, "Medicine: To Forecast Birth Defects," *New York Times,* May 25, 1969.

17. H. A, Lubs, M. Watson, R. Breg, E. Lujan, and J. M. Opitz, "Restudy of the Original Marker X Family," *American Journal of Medical Genetics* 17, no. 1 (1984): 133–44. The woman's sister, who was also a carrier of the marker X, had no children.

18. D. W. Smith, "Dysmorphology (Teratology)," *Journal of Pediatrics* 69, no. 6 (1966): 1150–69; V. A. McKusick, *Mendelian Inheritance in Man: Catalogs of Autosomal Dominant, Autosomal Recessive, and X-Linked Phenotypes,* 4th ed. (Baltimore: Johns Hopkins University Press, 1975).

19. V. A. McKusick, *Mendelian Inheritance in Man: Catalogs of Autosomal Dominant, Recessive, and X-Linked Phenotypes* (Baltimore: Johns Hopkins University Press, 1966). For more on McKusick and his cataloging of genetic disorders, see Susan Lindee, *Moments of Truth in Genetic Medicine* (Baltimore: Johns Hopkins University Press, 2005).

20. Smith, "Dysmorphology"; D. W. Smith, *Recognizable Patterns of Human Malformation* (Philadelphia: W. B. Saunders, 1970).

21. Smith, *Recognizable Patterns of Human Malformation.*

22. R. J. Gorlin and J. J. Pindborg, *Syndromes of the Head and Neck* (New York: McGraw-Hill, 1964); R. M. Goodman and R. J. Gorlin, *The Face in Genetic Disorders*

(St. Louis: C. V. Mosby, 1970). *Syndromes of the Head and Neck* has since gone through many editions, the most recent published in 2010. *The Face in Genetic Disorders* was also updated once, in 1977.

23. Goodman and Gorlin, *Face in Genetic Disorders.*

24. Lindee, *Moments of Truth,* 81.

25. J. M. Aase, "The Dysmorphology Detective," *Pediatric Annals* 10, no. 7 (1981): 38–43, 38.

26. David Wright, *Downs: The History of a Disability* (New York: Oxford University Press, 2011); Susan E. Lederer, *Subjected to Science: Human Experimentation in America before the Second World War* (Baltimore: Johns Hopkins University Press, 1994); David Wright, Laurie Jacklin, and Tom Themeles, "Dying to Get Out of the Asylum: Mortality and Madness in Four Mental Hospitals in Victorian Canada, c. 1841–1891," *Bulletin of the History of Medicine* 87, no. 4 (2013): 591–621; Kevles, *In the Name of Eugenics;* Theodore M. Porter, *The Data of Insanity: Asylum Statistics and the Investigation of Human Heredity since 1789* (Princeton, NJ: Princeton University Press, forthcoming).

27. Wright, *Downs;* Andrew J. Hogan, "Medical Eponyms: Patient Advocates, Professional Interests, and the Persistence of Honorary Naming," *Social History of Medicine,* forthcoming in 2016.

28. G. Turner, *Y the X? Unraveling Intellectual Disability and Autism* (Newcastle, NSW: Gillian Turner, 2012), 22.

29. Ibid., 24.

30. Ibid.

31. Ibid. According to Turner, Dunn wrote her a letter claiming that the disorder should in fact be called Dunn syndrome. As it turned out, the family Dunn described had fragile X syndrome, while the family Renpenning reported had a unique disorder still referred to as Renpenning syndrome. H. G. Dunn, H. Renpenning, J. W. Gerrard, J. R. Miller, T. Tabata, and S. Federoff, "Mental Retardation as a Sexlinked Defect," *American Journal of Mental Deficiency* 67, no. 6 (1963): 827–48; J. W. Gerrard and H. J. Renpenning, "Sex-Linked Mental Retardation," *Lancet* 303, no. 7870 (1974): 1346; H. Renpenning, J. W. Gerrard, W. A. Zaleski, and T. Tabata, "Familial Sex-Linked Mental Retardation," *Canadian Medical Association Journal* 87, no. 18 (1962): 954; G. Turner, B. Turner, and E. Collins, "Renpenning's Syndrome—X-Linked Mental Retardation," *Lancet* 2, no. 7668 (1970): 365.

32. G. Turner, B. Turner, and E. Collins, "X-Linked Mental Retardation without Physical Abnormality: Renpenning's Syndrome," *Developmental Medicine & Child Neurology* 13, no. 1 (1971): 71–78; G. Turner, B. Engisch, D. G. Lindsay, and B. Turner, "X-Linked Mental Retardation without Physical Abnormality (Renpenning's Syndrome) in Sibs in an Institution," *Journal of Medical Genetics* 9, no. 3 (1972): 324.

33. Kevles, *In the Name of Eugenics,* 159–60.

34. Penrose, *Outline of Human Genetics,* 83.

35. J. M. Opitz and G. R. Sutherland, "International Workshop on the Fragile X and X-Linked Mental Retardation," *American Journal of Medical Genetics* 17, no. 1 (1984): 5–94, 11.

36. A. Anastasi, "Four Hypotheses with a Dearth of Data: Response of Lehrke's 'A Theory of X-Linkage of Major Intellectual Traits,'" *American Journal of Mental Deficiency* 76, no. 6 (1972): 620–22; R. G. Lehrke, "A Theory of X-Linkage of Major Intellectual Traits," *American Journal of Mental Deficiency* (1972): 611–19; R. G. Lehrke, "X-Linked Mental Retardation and Verbal Disability," *Birth Defects Original Article Series* 10, no. 1 (1974); W. E. Nance and E. Engel, "One X and Four Hypotheses: Response to Lehkre's 'A Theory of X-Linkage of Major Intellectual Traits,'" *American Journal of Mental Deficiency* 76, no. 6 (1972): 623–25; G. Turner and J. M. Opitz, "X-Linked Mental Retardation," *American Journal of Medical Genetics* 7, no. 4 (1980): 407–15; J. M. Opitz, "Editorial Comment: On the Gates of Hell and a Most Unusual Gene," *American Journal of Medical Genetics* 23, no. 1 (1986): 1–10; Turner, *Y the X?*

37. Lehrke, "X-Linked Mental Retardation"; W. Allan, C. N. Herndon, and F. C. Dudley, "Some Examples of the Inheritance of Mental Deficiency: Apparently Sex-Linked Idiocy and Microcephaly," *American Journal of Mental Deficiency* 48 (1944): 325–34; J. P. Martin and J. Bell, "A Pedigree of Mental Defect Showing Sex-Linkage," *Journal of Neurological Psychiatry* 6, nos. 3–4 (1943): 154–57. For more on Allan, Herndon, and Dudley, see Nathaniel Comfort, *The Science of Human Perfection: Heredity and Health in American Biomedicine* (New Haven, CT: Yale University Press, 2012); Dunn et al., "Mental Retardation"; M. S. Losowsky, "Hereditary Mental Defect Showing the Pattern of Sex Influence," *Journal of Mental Deficiency Research* 5 (1961): 60–62; Renpenning et al., "Familial Sex-Linked Mental Retardation."

38. Lehrke, "X-Linked Mental Retardation."

39. Allan, Herndon, and Dudley, "Some Examples of Inheritance."

40. G. Turner, "Historical Overview of X-Linked Mental Retardation," in *The Fragile X Syndrome: Diagnosis, Biochemistry, and Intervention,* ed. R. J. Hagerman and P. M. McBogg (Dillon, CO: Spectra, 1983), 1–16; Turner, *Y the X?;* J. A. Escalante, H. Grunspun, and O. Frota-Pessoa, "Severe Sex-Linked Mental Retardation," *Journal de Genetique Humaine* 19 (1971): 137–40; Opitz and Sutherland, "International Workshop"; A. M. Vianna-Morgante, I. Armando, and O. Frota-Pessoa, "Escalante Syndrome and the Marker X Chromosome," *American Journal of Medical Genetics* 12, no. 2 (1982): 237–40.

41. A. Prader, "Testicular Size: Assessment and Clinical Importance," *Triangle* 7, no. 6 (1966): 240–43; M. Zachmann, A. Prader, H. P. Kind, H. Häfliger, and H. Budliger, "Testicular Volume during Adolescence: Cross-Sectional and Longitudinal Studies," *Helvetica Paediatrica Acta* 29, no. 1 (1974): 61–72; G. Turner, C. Eastman, J. Casey, A. McLeay, P. Procopis, and B. Turner, "X-Linked Mental Retardation Associated with Macro-orchidism," *Journal of Medical Genetics* 12, no. 4 (1975): 367–71; Turner, *Y the X?*

42. B. Biederman, P. Bowen, and K. Swallow, "Mental Retardation with Macroorchidism and Pedigree Consistent with X-Linked Inheritance," *Birth Defects* 13, no. 3C (1977): 224–25; J. M. Cantú, H. E. Scaglia, M. Medina, M. Gonzalez-Diddi, T. Morato, M. E. Moreno, and G. Perez-Palacios, "Inherited Congenital Normofunctional Testicular Hyperplasia and Mental Deficiency," *Human Genetics* 33, no. 1 (1976): 23–33; R. H. A. Ruvalcaba, S. A. Myhre, E. C. Roosen-Runge, and J. B. Beckwith, "X-Linked Mental Deficiency Megalotestes Syndrome," *Journal of the American Medical Association*

238, no. 15 (1977): 1646; P. Bowen, B. Biederman, K. A. Swallow, and J. M. Opitz, "The X-Linked Syndrome of Macroorchidism and Mental Retardation: Further Observations," *American Journal of Medical Genetics* 2, no. 4 (1978): 409–14, 412.

43. V. A. McKusick, *Mendelian Inheritance in Man: Catalogs of Autosomal Dominant, Autosomal Recessive, and X-Linked Phenotypes*, 5th ed. (Baltimore: Johns Hopkins University Press, 1978), 781; D. W. Smith and K. L. Jones, *Recognizable Patterns of Human Malformation: Genetic, Embryologic, and Clinical Aspects*, 3rd ed. (Philadelphia: W. B. Saunders, 1982), 120.

44. F. Giraud, S. Aymé, J. F. Mattei and M. G. Mattei, "Constitutional Chromosomal Breakage," *Human Genetics* 34, no. 2 (1976): 125–36; J. Harvey, C. Judge, and S. Wiener, "Familial X-Linked Mental Retardation with an X Chromosome Abnormality," *Journal of Medical Genetics* 14, no. 1 (1977): 46–50; Lubs, "Marker X Chromosome"; R. E. Magenis, F. Hecht, and E. W. Lovrien, "Heritable Fragile Site on Chromosome 16: Probably Localization of Haptoglobin Locus in Man," *Science* 170, no. 3953 (1970): 85–87.

45. Lindee, *Moments of Truth;* Löwy, "How Genetics Came"; Alexandra Minna Stern, *Telling Genes: The Story of Genetic Counseling in America* (Baltimore: Johns Hopkins University Press, 2012).

46. Ségolène Aymé, personal communication with author, August 20, 2012.

47. Giraud et al., "Constitutional Chromosomal Breakage"; Harvey, Judge, and Wiener, "Familial X-Linked Mental Retardation."

48. Turner, "Historical Overview."

49. Harvey, Judge, and Wiener, "Familial X-Linked Mental Retardation," 49–50.

50. G. R. Sutherland, "Fragile Sites on Human Chromosomes: Demonstration of Their Dependence on the Type of Tissue Culture Medium," *Science* 197, no. 4300 (1977): 265–66; Grant Sutherland, interview by Peter Harper, July 7, 2006, Interviews with Human and Medical Geneticists series, Special Collections and Archives, Cardiff University, Cardiff, UK.

51. Sutherland, "Fragile Sites on Human Chromosomes"; G. R. Sutherland, "Heritable Fragile Sites on Human Chromosomes: I, Factors Affecting Expression in Lymphocyte Culture," *American Journal of Human Genetics* 31, no. 2 (1979): 125–35; Sutherland, interview by Harper; P. S. Gerald, "X-Linked Mental Retardation and the Fragile-x Syndrome," *Pediatrics* 68, no. 4 (1981): 594–95.

52. G. R. Sutherland, "Heritable Fragile Sites on Human Chromosomes: III, Detection of Fra(x)(q27) in Males with X-Linked Mental Retardation and in Their Female Relatives," *Human Genetics* 53, no. 1 (1979): 23–27: G. Turner, R. Till, and A. Daniel, "Marker X Chromosomes, Mental Retardation and Macro-orchidism," *New England Journal of Medicine* 299, no. 26 (1978): 1472; Turner, *Y the X?*

53. P. N. Howard-Peebles and G. R. Stoddard, "Familial X-Linked Mental Retardation with a Marker X Chromosome and Its Relationship to Macro-orchidism," *Clinical Genetics* 17, no. 2 (1980): 125–28; P. A. Jacobs, T. W. Glover, M. Mayer, P. Fox, J. W. Gerrard, H. G. Dunn, D. S. Herbst, and J. M. Optiz, "X-Linked Mental Retardation: A Study of 7 Families," *American Journal of Medical Genetics* 7, no. 4 (1980): 471–89; G. R. Sutherland and P. L. C. Ashforth, "X-Linked Mental Retardation with Macro-orchidism and the Fragile Site at Xq27 or 28," *Human Genetics* 48, no. 1 (1979): 117–20; G. Turner, R. Brookwell,

A. Daniel, M. Selikowitz, and M. Zilibowitz, "Heterozygous Expression of X-Linked Mental Retardation and X-Chromosome Marker Fra (x)(q27)," *New England Journal of Medicine* 303, no. 12 (1980): 662–64.

54. McKusick, *Mendelian Inheritance in Man* (1978), xviii.

55. Adam M. Hedgecoe, "Expansion and Uncertainty: Cystic Fibrosis, Classification and Genetics," *Sociology of Health & Illness* 25, no. 1 (2003): 50–70, 58.

56. Turner et al., "Heterozygous Expression"; Turner, "Historical Overview"; Turner, *Y the X?;* Andrew J. Hogan, "Visualizing Carrier Status: Fragile X Syndrome and Genetic Diagnosis since the 1940s," *Endeavour* 36, no. 2 (2012): 77–84.

57. F. C. Fraser, "Of Mice and Children: Reminiscences of a Teratogeneticist," *American Journal of Medical Genetics Part A* 146A, no. 17 (2008): 2179–2202; 2188; Löwy, "How Genetics Came"; Stern, *Telling Genes*.

58. McKusick, *Mendelian Inheritance in Man* (1978), 780.

59. Penrose, *Outline of Human Genetics,* 136.

60. Lubs, "Marker X Chromosome"; Gerald, "Mental Retardation and Fragile-x Syndrome"; P. S. Gerald, "X-Linked Mental Retardation and an X-Chromosome Marker," *New England Journal of Medicine* 303, no. 12 (1980): 696–97; V. A. McKusick, "The Human Genome through the Eyes of Mercator and Vesalius," *Transactions of the American Clinical and Climatological Association* 92 (1981): 66–90; C. R. Scriver, "Window Panes of Eternity: Health, Disease, and Inherited Risk," *Yale Journal of Biology and Medicine* 55, nos. 5–6 (1982): 487–513.

61. Lubs, "Marker X Chromosome"; Turner, Till, and Daniel, "Marker X Chromosomes"; Löwy, "How Genetics Came."

62. Turner et al., "Heterozygous Expression."

63. Ruth Schwartz Cowan, *Heredity and Hope: The Case for Genetic Screening* (Cambridge, MA: Harvard University Press, 2008).

64. Sutherland, "Heritable Fragile Sites: III"; Jacobs et al., "X-Linked Mental Retardation"; Howard-Peebles and Stoddard, "Familial X-Linked Mental Retardation."

65. Sutherland, "Heritable Fragile Sites III," 26; P. B. Jacky and F. J. Dill, "Expression in Fibroblast Culture of the Satellited-x Chromosome Associated with Familial Sex-Linked Mental Retardation," *Human Genetics* 53, no. 2 (1980): 267–69; E. C. Jenkins, W. T. Brown, C. J. Duncan, J. Brooks, M. Ben-Yishay, F. M. Giordano, and H. M. Nitowsky, "Feasibility of Fragile X Chromosome Prenatal Diagnosis Demonstrated," *Lancet* 2, no. 8258 (1981): 1292; Edmund Jenkins, interview by author, New York, May 26, 2011.

66. Randi Hagerman, interview by author, Sacramento, CA, March 2, 2012; Loris McGavran, interview by author, Denver, August 20, 2012.

67. Hagerman and McBogg, *Fragile X Syndrome;* Martin and Bell, "Pedigree of Mental Defect"; V. A. McKusick, *Mendelian Inheritance in Man: Catalogs of Autosomal Dominant, Recessive, and X-Linked Phenotypes,* 6th ed. (Baltimore: Johns Hopkins University Press, 1983), 1070.

68. Hagerman interview.

69. Opitz, "Editorial Comment."

70. S. L. Sherman, N. E. Morton, P. A. Jacobs, and G. Turner, "The Marker (x) Syndrome: A Cytogenetic and Genetic Analysis," *Annals of Human Genetics* 48, no. 1 (1984):

21–37; M. E. Pembrey, R. M. Winter, K. E. Davies, J. M. Opitz, and J. F. Reynolds, "A Premutation That Generates a Defect at Crossing Over Explains the Inheritance of Fragile X Mental Retardation," *American Journal of Medical Genetics* 21, no. 4 (1985): 709–17.

71. J. Donachie and D. G. Monckton, "Stephanie L. Sherman Interview," *Tomorrow Belongs to Me* (Glasgow, UK: University of Glasgow Press, 2006), 155; Lindee, *Moments of Truth*.

72. Sherman et al., "Marker (x) Syndrome"; S. L. Sherman, P. A. Jacobs, N. E. Morton, U. Froster-Iskenius, P. N. Howard-Peebles, K. B. Nielsen, M. W. Partington, G. R. Sutherland, G. Turner, and M. Watson, "Further Segregation Analysis of the Fragile X Syndrome with Special Reference to Transmitting Males," *Human Genetics* 69, no. 4 (1985): 289–99; Pembrey et al., "Premutation That Generates a Defect"; W. T. Brown, E. C. Jenkins, M. S. Krawczun, K. Wisniewski, R. Rudelli, I. L. Cohen, G. Fisch, E. Wolf-Schein, C. Miezejeski, and C. Dobkin, "The Fragile X Syndrome," *Annals of the New York Academy of Sciences* 477, no. 1 (1986): 129–50. On anticipation, see Judith E. Friedman, "Anticipation in Hereditary Disease: The History of a Biomedical Concept," *Human Genetics* (2011): 1–10.

73. Donachie and Monckton, "Sherman Interview."

74. G. Turner, J. M. Opitz, W. T. Brown, K. E. Davies, P. A. Jacobs, E. C. Jenkins, M. Mikkelsen, M. W. Partington, G. R. Sutherland, and J. F. Reynolds, "Second International Workshop on the Fragile X and on X-Linked Mental Retardation," *American Journal of Medical Genetics* 23, nos. 1–2 (1986): 11–67, 56; Donachie and Monckton, "Sherman Interview," 156.

75. Opitz, "Editorial Comment," 8.

76. S. F. Hoegerman, J. M. Rary, J. M. Opitz, and J. F. Reynolds, "Speculation on the Role of Transposable Elements in Human Genetic Disease with Particular Attention to Achondroplasia and the Fragile X Syndrome," *American Journal of Medical Genetics* 23, nos. 1–2 (1986): 685–99; J. M. Friedman, P. N. Howard-Peebles, J. M. Opitz, and J. F. Reynolds, "Inheritance of Fragile X Syndrome: An Hypothesis," *American Journal of Medical Genetics* 23, nos. 1–2 (1986): 701–13.

77. G. R. Sutherland, E. Baker, J. M. Opitz, and J. F. Reynolds, "Effects of Nucleotides on Expression of the Folate Sensitive Fragile Sites," *American Journal of Medical Genetics* 23, nos. 1–2 (1986): 409–17.

78. R. L. Nussbaum and D. H. Ledbetter, "Fragile X Syndrome: A Unique Mutation in Man," *Annual Review of Genetics* 20, no. 1 (1986): 109–41; D. H. Ledbetter, S. A. Ledbetter, and R. L. Nussbaum, "Implications of Fragile X Expression in Normal Males for the Nature of the Mutation," *Nature* 324, no. 6093 (1986): 161–63.

79. Pembrey et al., "Premutation That Generates a Defect." For more on these models, see Turner et al., "Second International Workshop"; K. E. Davies, *The Fragile X Syndrome* (New York: Oxford University Press, 1989).

80. Andrew J. Hogan, "Disrupting Genetic Dogma: Bridging Cytogenetics and Molecular Biology in Fragile X Research," *Historical Studies of the Natural Sciences* 45, no. 1 (2015): 174–97.

81. W. T. Brown, A. Gross, C. Chan, E. C. Jenkins, J. L. Mandel, I. Oberlé, B. Arveiler, G. Novelli, S. Thibodeau, and R. Hagerman, "Multilocus Analysis of the Fragile X

Syndrome," *Human Genetics* 78, no. 3 (1988): 201–5; N. Dahl, P. Goonewardena, H. Malmgren, K. H. Gustavson, G. Holmgren, E. Seemanova, G. Annerén, A. Flood, and U. Pettersson, "Linkage Analysis of Families with Fragile-x Mental Retardation, Using a Novel RFLP Marker (dxs 304)," *American Journal of Human Genetics* 45, no. 2 (1989): 304; R. Heilig, I. Oberlé, B. Arveiler, A. Hanauer, M. Vidaud, and J. L. Mandel, "Improved DNA Markers for Efficient Analysis of Fragile X Families," *American Journal of Medical Genetics* 30, nos. 1–2 (1988): 543–50; C. Nguyen, M. G. Mattei, J. A. Rey, M. A. Baeteman, J. F. Mattei, and B. R. Jordan, "Cytogenetic and Physical Mapping in the Region of the X Chromosome Surrounding the Fragile Site," *American Journal of Medical Genetics* 30, nos. 1–2 (1988): 601–11; B. A. Oostra, P. E. Hupkes, L. F. Perdon, C. A. Van Bennekom, E. Bakker, D. J. J. Halley, M. Schmidt, D. Du Sart, A. Smits, and B. Wieringa, "New Polymorphic DNA Marker Close to the Fragile Site Fraxa," *Genomics* 6, no. 1 (1990): 129–32.

82. Hagerman interview.

83. Y. H. Fu, D. Kuhl, A. Pizzuti, M. Pieretti, J. S. Sutcliffe, S. Richards, A. J. M. H. Verkerk, J. J. A. Holden, and R. G. Fenwick, "Variation of the CGG Repeat at the Fragile X Site Results in Genetic Instability: Resolution of the Sherman Paradox," *Cell* 67, no. 6 (1991): 1047–58; A. J. M. H. Verkerk, M. Pieretti, J. S. Sutcliffe, Y. H. Fu, D. Kuhl, A. Pizzuti, O. Reiner, S. Richards, M. F. Victoria, and F. Zhang, "Identification of a Gene (fmr-1) Containing a Cgg Repeat Coincident with a Breakpoint Cluster Region Exhibiting Length Variation in Fragile X Syndrome," *Cell* 65, no. 5 (1991): 905–14; A. Vincent, D. Hertz, C. Petit, C. Kretz, I. Oberlé, and J. L. Mandel, "Abnormal Pattern Detected in Fragile-x Patients by Pulsed-Field Gel Electrophoresis," *Nature* 349 (1991): 624–26; S. Yu, J. Mulley, D. Loesch, G. Turner, A. Donnelly, A. Gedeon, D. Hillen, E. Kremer, M. Lynch, and M. Pritchard, "Fragile-x Syndrome: Unique Genetics of the Heritable Unstable Element," *American Journal of Human Genetics* 50, no. 5 (1992): 968–80.

84. I. Oberle, F. Rousseau, D. Heitz, and C. Kretz, "Instability of a 550-Base Pair DNA Segment and Abnormal Methylation in Fragile X Syndrome," *Science* 252, no. 5009 (1991): 1097–1102; S. L. Nolin, F. A. Lewis, L. L. Ye, G. E. Houck, L. E. Glicksman, P. Limprasert, S. Y. Li, N. Zhong, A. E. Ashley, E. Feingold, S. L. Sherman, and W. T. Brown, "Familial Transmission of the Fmr1 CGG Repeat," *American Journal of Human Genetics* 59, no. 6 (1996): 1252–61.

85. E. C. Jenkins, G. E. Houck, X. H. Ding, S. Y. Li, S. L. Stark-Houck, J. Salerno, M. Genovese, A. Glicksman, S. L. Nolin, and N. Zhong, "An Update on Fragile X Prenatal Diagnosis: End of the Cytogenetic Testing Era," *Developmental Brain Dysfunction* 8 (1995): 293–301.

86. D. J. Allingham-Hawkins, R. Babul-Hirji, D. Chitayat, J. J. A. Holden, K. T. Yang, C. Lee, R. Hudson, H. Gorwill, S. L. Nolin, and A. Glicksman, "Fragile X Premutation Is a Significant Risk Factor for Premature Ovarian Failure: The International Collaborative POF in Fragile X Study—Preliminary Data," *American Journal of Medical Genetics* 83, no. 4 (1999): 322–25; P. J. Hagerman and R. J. Hagerman, "The Fragile-x Premutation: A Maturing Perspective," *American Journal of Human Genetics* 74, no. 5 (2004): 805–16; Hogan, "Visualizing Carrier Status"; Lindee, *Moments of Truth;* Donachie and Monckton, "Sherman Interview."

87. P. S. Harper, H. G. Harley, W. Reardon, and D. J. Shaw, "Anticipation in Myotonic Dystrophy: New Light on an Old Problem," *American Journal of Human Genetics* 51, no. 1 (1992): 10–16; G. R. Sutherland, E. Kremer, M. Lynch, M. Pritchard, S. Yu, R. I. Richards, and E. A. Haan, "Hereditary Unstable DNA: A New Explanation for Some Old Genetic Questions?," *Lancet* 338, no. 8762 (1991): 289–92; Friedman, "Anticipation in Hereditary Disease."

88. M. E. MacDonald, C. M. Ambrose, M. P. Duyao, R. H. Myers, C. Lin, L. Srinidhi, G. Barnes, S. A. Taylor, M. James, and N. Groot, "A Novel Gene Containing a Trinucleotide Repeat That Is Expanded and Unstable on Huntington's Disease Chromosomes," *Cell* 72, no. 6 (1993): 971–83; H. T. Orr, M. Chung, S. Banfi, T. J. Kwiatkowski, A. Servadio, A. L. Beaudet, A. E. McCall, L. A. Duvick, L. P. W. Ranum, and H. Y. Zoghbi, "Expansion of an Unstable Trinucleotide CAG Repeat in Spinocerebellar Ataxia Type 1," *Nature Genetics* 4, no. 3 (1993): 221–26.

89. Peter Keating and Alberto Cambrosio, "The New Genetics and Cancer: The Contributions of Clinical Medicine in the Era of Biomedicine," *Journal of the History of Medicine and Allied Sciences* 56, no. 4 (2001): 321–52.

90. H. A. Lubs and F. de la Cruz, eds., *Genetic Counseling* (New York: Raven Press, 1977), v.

91. Comfort, *Science of Human Perfection*.

92. Lubs, "Marker X Chromosome"; Scriver, "Window Panes of Eternity."

93. P. Berg, "Dissections and Reconstructions of Genes and Chromosomes," *Bioscience Reports* 1, no. 4 (1981): 269–87; V. A. McKusick, "Birth Defects—Prospects for Progress," in *Congenital Malformations: Proceedings of the Third International Conference,* ed. F. Clarke Fraser and V. A. McKusick (Amsterdam: Excerpta Medica, 1970), 407–13; B. Glass, "Science: Endless Horizons or Golden Age?," *Science* 171, no. 3966 (1971): 23–29; McKusick, "Human Genome through the Eyes"; A. G. Motulsky, "Presidential Address: Human and Medical Genetics: Past, Present, and Future," in *Human Genetics: Proceedings of the 7th International Conference, Berlin 1986* (Berlin: Springer-Verlag, 1987), 3–13. As historian of medicine Charles Rosenberg insightfully described, "enchantment" with the promise of gene therapy continued into the first decade of the twenty-first century. Charles E. Rosenberg, *Our Present Complaint: American Medicine, Then and Now* (Baltimore: Johns Hopkins University Press, 2007), 96–112.

Chapter 2. Chromosomal Cartography

1. V. A. McKusick, "Birth Defects—Prospects for Progress," in *Congenital Malformations: Proceedings of the Third International Conference,* ed. F. Clarke Fraser, and V. A. McKusick (Amsterdam: Excerpta Medica, 1970), 407–13, 407–8.

2. Peter Galison and Bruce William Hevly, *Big Science: The Growth of Large Scale Research* (Stanford, CA: Stanford University Press, 1992); Angela N. H. Creager, *Life Atomic: A History of Radioisotopes in Science and Medicine* (Chicago: University of Chicago Press, 2013); Daniel J. Kevles, *The Physicists: The History of a Scientific Community in Modern America* (Cambridge, MA: Harvard University Press, 1971); Toby A. Appel, *Shaping Biology: The National Science Foundation and American Biology Research, 1945–1975* (Balti-

more: Johns Hopkins University Press, 2003); Robert Cook-Deegan, *Gene Wars: Science, Politics, and the Human Genome* (New York: Norton, 1994).

3. V. A. McKusick, *Mendelian Inheritance in Man: Catalogs of Autosomal Dominant, Autosomal Recessive, and X-Linked Phenotypes,* 3rd ed. (Baltimore: Johns Hopkins University Press, 1971), xiii; V. A. McKusick, *Mendelian Inheritance in Man: Catalogs of Autosomal Dominant, Autosomal Recessive, and X-Linked Phenotypes,* 5th ed. (Baltimore: Johns Hopkins University Press, 1978), xx.

4. Appel, *Shaping Biology,* 235–36.

5. Reed Pyeritz, interview by author, Philadelphia, April 18, 2012.

6. D. G. Harnden, "Early Studies on Human Chromosomes," *BioEssays* 18, no. 2 (1996): 163–68; J. D. Rowley, "Theodore T. Puck (September 24, 1916–November 6, 2005)," *American Journal of Human Genetics* 78, no. 3 (2006): 365–66; Susan Lindee, *Moments of Truth in Genetic Medicine* (Baltimore: Johns Hopkins University Press, 2005).

7. Denver Study Group, "A Proposed Standard System of Nomenclature of Human Mitotic Chromosomes," *Lancet* 275, no. 7133 (1960): 1063–65; London Conference on the Normal Human Karyotype, *Cytogenetics* 2 (1963): 264–68.

8. Chicago Conference, "Standardization in Human Cytogenetics," *Birth Defects Original Article Series* 2, no. 2 (1966).

9. Uta Francke, interview by author, Palo Alto, CA, February 27, 2012; Kurt Hirschhorn, interview by author, New York, January 26, 2012; Dorothy Warburton, telephone interview by author, May 11, 2011.

10. Hirschhorn interview. Peter Harper offered a similar account of how "p+q=1" came about during his interview of David Harden. David Harnden, interview by Peter Harper, March 18, 2004, Interviews with Human and Medical Geneticists series, Special Collections and Archives, Cardiff University, Cardiff, UK, 12.

11. H. A. Lubs, "A Marker X Chromosome," *American Journal of Human Genetics* 21, no. 3 (1969): 231–44.

12. V. A. McKusick, "A History of Medical Genetics," in *Emery and Rimoin's Principles and Practice of Medical Genetics* (New York: Elsevier Health Sciences, 1997), 1–30, 8.

13. T. Caspersson, L. Zech, C. Johansson, and E. J. Modest, "Identification of Human Chromosomes by DNA-Binding Fluorescent Agents," *Chromosoma* 30, no. 2 (1970): 215–27; M. Seabright, "A Rapid Banding Technique for Human Chromosomes," *Lancet* 2, no. 7731 (1971): 971–72.

14. Warburton interview.

15. Paris Conference (1971), "Standardization in Human Cytogenetics," *Birth Defects Original Article Series* 8, no. 7 (1972): 1–46.

16. V. A. McKusick, "A 60-Year Tale of Spots, Maps, and Genes," *Annual Reviews of Genomics and Human Genetics* 7 (2006): 1–27.

17. Robert E. Kohler, *Lords of the Fly: Drosophila Genetics and the Experimental Life* (Chicago: University of Chicago Press, 1994); Raphael Falk, "Linkage: From Particulate to Interactive Genetics," *Journal of the History of Biology* 36, no. 1 (2003): 87–117; Seabright, "Rapid Banding Technique."

18. J. B. S. Haldane, "Some Alternatives to Sex," *New Biology* 19 (1955): 7–26; V. A. McKusick and F. H. Ruddle, "The Status of the Gene Map of the Human Chromosomes," *Science* 196, no. 4288 (1977): 390–405; Nathaniel Comfort, *The Science of Human Perfection: Heredity and Health in American Biomedicine* (New Haven, CT: Yale University Press, 2012).

19. H. Tijo and A. Levan, "The Chromosomes of Man," *Hereditas* 42, no. 1 (1956): 1–6; G. Barski, S. Sorieul, and F. Cornefert, "Production of Cells of a 'Hybrid' Nature in Cultures in vitro of 2 Cellular Strains in Combination," *Comptes rendus hebdomadaires des seances de l'Academie des Sciences* 251 (1960): 1825–27; J. W. Littlefield, "Selection of Hybrids from Matings of Fibroblasts in vitro and Their Presumed Recombinants," *Science* 145, no. 3633 (1964): 709–10.

20. Hannah Landecker, *Culturing Life: How Cells Became Technologies* (Cambridge, MA: Harvard University Press, 2007), 180–218.

21. H. Harris and J. F. Watkins, "Hybrid Cells Derived from Mouse and Man: Artificial Heterokaryons of Mammalian Cells from Different Species," *Nature* 205 (1965): 640–46; M. C. Weiss and H. Green, "Human-Mouse Hybrid Cell Lines Containing Partial Complements of Human Chromosomes and Functioning Human Genes," *Proceedings of the National Academy of Sciences of the United States of America* 58, no. 3 (1967): 1104–11.

22. Weiss and Green, "Human-Mouse Hybrid Cell Lines," 1111.

23. Ibid.; B. R. Migeon and C. S. Miller, "Human-Mouse Somatic Cell Hybrids with Single Human Chromosome (Group E): Link with Thymidine Kinase Activity," *Science* 162, no. 3857 (1968): 1005–6.

24. McKusick, "Birth Defects," 408.

25. C. Boone, T. Chen, and F. H. Ruddle, "Assignment of Three Human Genes to Chromosomes (ldh-a to 11, Tk to 17, and Idh to 20) and Evidence for Translocation between Human and Mouse Chromosomes in Somatic Cell Hybrids," *Proceedings of the National Academy of Sciences* 69, no. 2 (1972): 510–14.

26. S. M. Elsevier, R. S. Kucherlapati, E. A. Nichols, R. P. Creagan, R. E. Giles, F. H. Ruddle, K. Willecke, and J. K. McDougall, "Assignment of the Gene for Galactokinase to Human Chromosome 17 and Its Regional Localisation to Band Q21–22," *Nature* 251 (1974): 635–36.

27. F. H. Ruddle, "Linkage Analysis in Man by Somatic Cell Genetics," *Nature* 242 (1973): 165–69.

28. Lewis L. Coriell, "Cell Repository," *Science* 180, no. 4084 (1973): 427; A. E. Greene, R. C. Miller, and L. L. Coriell, "Appendix 2: Cell Cultures with Chromosomal Aberrations—July 1974—Available from the Human Genetic Mutant Cell Repository," *Cytogenetic and Genome Research* 14, nos. 3–6 (1975): 470–75; A. E. Greene, R. A. Mulivor, and L. L. Coriell, "The Role of the Human Genetic Mutant-Cell Repository in the Mapping of Human Genes," *Cytogenetics and Cell Genetics* 25, nos. 1–4 (1979): 162.

29. A. de la Chapelle, R. C. Miller, A. E. Greene, and L. L. Coriell, "A (1; 17) Translocation, Balanced, Plus Trisomy 21, 47 Chromosomes," *Cytogenetic and Genome Research* 14, no. 1 (1975): 82–83.

30. *The Human Genetic Mutant Cell Repository: List of Genetic Variants, Chromosomal Aberrations and Normal Cell Cultures Submitted to the Repository, Fourth Edition, October 1977* (Bethesda: US Department of Health, Education, and Welfare, 1977), 22.

31. Peter Keating and Alberto Cambrosio, *Biomedical Platforms: Realigning the Normal and the Pathological in Late-Twentieth Century Medicine* (Cambridge, MA: MIT Press, 2003); Peter Keating and Alberto Cambrosio, "The New Genetics and Cancer: The Contributions of Clinical Medicine in the Era of Biomedicine," *Journal of the History of Medicine and Allied Sciences* 56, no. 4 (2001): 321–52; Ilana Löwy, *Between Bench and Bedside: Science, Healing, and Interleukin-2 in a Cancer Ward* (Cambridge, MA: Harvard University Press, 1996); Soraya de Chadarevian and Harmke Kamminga, eds., *Molecularizing Biology and Medicine: New Practices and Alliances, 1910s–1970s* (Amsterdam: Harwood Academic, 1998).

32. Francke interview.

33. Ibid.; U. Francke, "2012 William Allan Award: Adventures in Cytogenetics," *American Journal of Human Genetics* 92, no. 3 (2013): 325–37; Appel, *Shaping Biology.*

34. McKusick, "60-Year Tale"; V. A. McKusick, "The Morbid Anatomy of the Human Genome: A Review of Gene Mapping in Clinical Medicine (First of Four Parts)," *Medicine* 65, no. 1 (1986): 1–33; "New Haven Conference (1973): First International Workshop on Human Gene Mapping," *Birth Defects Original Article Series* 10, no. 3 (1974): 1–216.

35. "New Haven Conference (1973)"; "Rotterdam Conference (1974): Second International Workshop on Human Gene Mapping," *Birth Defects Original Article Series* 11, no. 3 (1975): 1–310.

36. "Baltimore Conference (1975): Third International Workshop on Human Gene Mapping," *Birth Defects Original Article Series* 12, no. 6 (1976): 1–452; "Human Gene Mapping 4 (1977): Fourth International Workshop on Human Gene Mapping," *Birth Defects Original Article Series* 14, no. 4 (1978): 1–730.

37. Francke, "2012 William Allan Award"; V. A. McKusick, "The Human Genome through the Eyes of Mercator and Vesalius," *Transactions of the American Clinical and Climatological Association* 92 (1981): 66–90, 77.

38. McKusick and Ruddle, "Status of the Gene Map."

39. Francke, "2012 William Allan Award," 326.

40. Harold M. Schmeck Jr., "Scientists Now Can Make Human Genetic Maps," *New York Times,* September 14, 1975, Week in Review, 188.

41. McKusick and Ruddle, "Status of the Gene Map," 401.

42. McKusick, "Human Genome through the Eyes," 66; McKusick, "Birth Defects," 408.

43. F. H. Ruddle, "The William Allan Memorial Award Address: Reverse Genetics and Beyond," *American Journal of Human Genetics* 36, no. 5 (1984): 944–53, 946.

44. David J. Rothman, *Strangers at the Bedside: A History of How Law and Bioethics Transformed Medical Decision Making* (New York: Basic Books, 1991); Susan Wright, *Molecular Politics: Developing American and British Regulatory Policy for Genetic Engineering, 1972–1982* (Chicago: University of Chicago Press, 1994); Susan M. Reverby, *Examining Tuskegee: The Infamous Syphilis Study and Its Legacy* (Chapel Hill: University of North Carolina Press, 2013); Comfort, *Science of Human Perfection,* 221–25.

45. Abby Lippman, "Led (Astray) by Genetic Maps: The Cartography of the Human Genome and Health Care," *Social Science & Medicine* 35, no. 12 (1992): 1469–76, 1474–75.

46. V. A. McKusick, "The Anatomy of the Human Genome," *Journal of Heredity* 71, no. 6 (1980): 370–91, 370.

47. Lindee, *Moments of Truth;* A. Chakravarti, "Obituary: Victor Almon McKusick (1921–2008)," *Nature* 455, no. 7209 (2008): 46.

48. V. A. McKusick, "The Anatomy of the Human Genome," *Hospital Practice* 16, no. 4 (1981): 82–100, 82.

49. Andrea Carlino, "Feature Review: Andreas Vesalius, *On the Fabric of the Human Body*," *Isis* 92, no. 1 (2001): 126–27.

50. Katharine Park, *Secrets of Women: Gender, Generation, and the Origins of Human Dissection* (New York: Zone Books, 2006), 21.

51. C. R. Scriver, "Window Panes of Eternity: Health, Disease, and Inherited Risk," *Yale Journal of Biology and Medicine* 55, nos. 5–6 (1982): 487–513, 496.

52. Cook-Deegan, *Gene Wars,* 120; Charles Scriver, telephone interview by author, May 30, 2012.

53. V. A. McKusick, "The Growth and Development of Human Genetics as a Clinical Discipline," *American Journal of Human Genetics* 27, no. 3 (1975): 261–73, 271.

54. David Gugerli, "Mapping: A Communicative Strategy," in *From Molecular Genetics to Genomics: The Mapping Cultures of Twentieth-Century Genetics,* ed. Jean-Paul Gaudillière and Hans-Jorg Rheinberger (London: Routledge, 2004), 210–18; Benedict Anderson, *Imagined Communities: Reflections on the Origin and Spread of Nationalism* (London: Verso, 1983).

55. Scriver interview.

56. Daniel J. Kevles, *In the Name of Eugenics: Genetics and the Uses of Human Heredity* (New York: Alfred A. Knopf, 1985).

57. Lindee, *Moments of Truth;* Ruth Schwartz Cowan, *Heredity and Hope: The Case for Genetic Screening* (Cambridge, MA: Harvard University Press, 2008); Diane B. Paul, *The Politics of Heredity: Essays on Eugenics, Biomedicine, and the Nature-Nurture Debate* (Albany: State University of New York Press, 1998); Diane B. Paul and Jeffrey P. Brosco, *The PKU Paradox: A Short History of a Genetic Disease* (Baltimore: Johns Hopkins University Press, 2013).

58. For differing perspectives on these genetic disease prevention programs and their larger social implications, see Cowan, *Heredity and Hope;* Keith Wailoo and Stephen Pemberton, *The Troubled Dream of Genetic Medicine: Ethnicity and Innovation in Tay-Sachs, Cystic Fibrosis, and Sickle Cell Disease* (Baltimore: Johns Hopkins University Press, 2006).

59. A. Milunsky, "Prenatal Diagnosis of Genetic Disorders," *American Journal of Medicine* 70, no. 1 (1981): 7–8, 7.

60. P. C. Nowell and D. A. Hungerford, "Chromosome Studies on Normal and Leukemic Human Leukocytes," *Journal of the National Cancer Institute* 25, no. 1 (1960): 85–109; G. Levan and F. Mitelman, "Clustering of Aberrations to Specific Chromosomes in Human Neoplasms," *Hereditas* 79, no. 1 (1975): 156–60; J. D. Rowley, "A New Consistent Chromosomal Abnormality in Chronic Myelogenous Leukemia Identified by Quinacrine Fluorescence and Giemsa Staining," *Nature* 243 (1973): 290–93.

61. Harold M. Schmeck Jr., "Progress Cited in Heredity Research," *New York Times,* August 21, 1975, 20.

62. A. Milunsky, *Know Your Genes* (Boston: Houghton Mifflin, 1977), 307.

63. T. Ferrell and V. Adams, "Ideas & Trends in Summary: Counting Up the Genes," *New York Times,* April 24, 1977, Week in Review, E6.

64. McKusick, "Human Genome through the Eyes," 76.

65. McKusick, "Anatomy of the Human Genome" (1981), 100.

66. T. Friedmann, *Gene Therapy: Fact and Fiction in Biology's New Approaches to Disease* (Cold Spring Harbor, NY: Cold Spring Harbor Laboratory, 1983), 75.

67. Ibid., 116.

68. McKusick, "Birth Defects," 408.

69. Milunsky, *Know Your Genes,* 1; H. J. Muller, "Our Load of Mutations," *American Journal of Human Genetics* 2, no. 2 (1950): 111–76; Robert W. Stock, "Will the Baby Be Normal?," *New York Times,* March 23, 1969, Sunday Magazine, 27; Lawrence K. Altman, "Genetic Help for the Laymen," *New York Times,* October 8, 1977, Sports, 21; Maya Pines, "Heredity Insurance," *New York Times,* April 30, 1978, Sunday Magazine, 78.

70. Milunsky, *Know Your Genes,* 1.

71. Cook-Deegan, *Gene Wars.*

72. V. A. McKusick, "The Gene Map of Homo Sapiens: Status and Prospectus," *Cold Spring Harbor Symposia on Quantitative Biology* 51 (1986): 15–27; McKusick, "60-Year Tale of Spots," 15.

73. Francke, "2012 William Allan Award," 328.

Chapter 3. The Genome's Morbid Anatomy

1. For additional analysis of presymptomatic diagnosis in medical genetics outside of the prenatal context, see Diane B. Paul and Jeffrey P. Brosco, *The PKU Paradox: A Short History of a Genetic Disease* (Baltimore: Johns Hopkins University Press, 2013); Stefan Timmermans and Mara Buchbinder, "Patients-in-Waiting: Living between Sickness and Health in the Genomics Era," *Journal of Health and Social Behavior* 51, no. 4 (2010): 408–23.

2. "Dr. Andrea Prader," *Gathered View* 37, no. 6 (2012): 1–2; Merlin Butler, Phillip D. K. Lee, and Barbara Y. Whitman, *Management of Prader-Willi Syndrome,* 3rd ed. (New York: Springer, 2006), "Appendix A: First Published Report of Prader-Willi Syndrome," 467–72; Christen Haigh, "Collaboration of Physicians behind First Identification of Prader-Willi Syndrome," *Endocrine Today,* October 2008.

3. Haigh, "Collaboration of Physicians"; "Dr. Andrea Prader"; Peter Beighton and Greta Beighton, *The Man behind the Syndrome* (Berlin: Springer-Verlag, 1986), 57; H. R. Wiedemann, "Guido Fanconi (1892–1979) in Memoriam," *European Journal of Pediatrics* 132, no. 3 (1979): 131–32.

4. H. R. Wiedemann, "Andrea Prader: On the Occasion of His 65th Birthday," *European Journal of Pediatrics* 143 (1984): 80–81; M. Zachmann, "Andrea Prader, 1919–2001," *Hormone Research in Paediatrics* (2002): 205–7; H. R. Wiedemann, "David W. Smith (1926–1981)," *European Journal of Pediatrics* 150 (1991): 533.

5. A. Prader, A. Labhart, H. Willi, and G. Fanconi, "Ein Syndrom von Adipositas, Kleinwuchs, Kryptorchismus und Idiotie bei Kindern und Erwachsenen, die als Neuge-

borene ein myatonie-artiges Bild geboten haben," *Proceedings of the Eighth International Congress of Pediatrics, Copenhagen* (1956): 13.

6. H. Zellweger and H. J. Schneider, "Syndrome of Hypotonia-Hypomentia-Hypogonadism-Obesity (HHHO) or Prader-Willi Syndrome," *Archives of Pediatrics and Adolescent Medicine* 115, no. 5 (1968): 588–98, 595–96.

7. Butler, Lee, and Whitman, *Management of Prader-Willi Syndrome,* ii; N. Bukvic and J. W. Elling, "Genetics in the Art and Art in Genetics," *Gene* 555, no. 1 (2015): 14–22.

8. O. C. Ward, "Down's 1864 Case of Prader-Willi Syndrome: A Follow-Up Report," *Journal of the Royal Society of Medicine* 90, no. 12 (1997): 694–69.

9. H. Bruch, "The Fröhlich Syndrome: Report of the Original Case," *American Journal of Diseases of Children* 58, no. 6 (1939): 1282–89.

10. Ibid., 1286. Bruch was quoting from Arthur Biedl, *Innere Sekretion: Ihre physiologischen Grundlagen und ihre Bedeutung für die Pathologie,* 2nd ed. (Berlin: Urban & Schwarzenberg, 1913); Beighton and Beighton, *Man behind the Syndrome,* 17.

11. Steven J. Peitzman, "From Dropsy to Bright's Disease to End-Stage Renal Disease," *Milbank Quarterly* 67 (1989): 16–32; Andrew J. Hogan, "Medical Eponyms: Patient Advocates, Professional Interests, and the Persistence of Honorary Naming," *Social History of Medicine,* forthcoming in 2016.

12. O. H. Wolff, "Obesity in Childhood: A Study of the Birth Weight, the Height, and the Onset of Puberty," *Quarterly Journal of Medicine* 24, no. 2 (1955): 109–23; Bruch, "Fröhlich Syndrome."

13. A. P. Schachat and I. H. Maumenee, "Bardet-Biedl Syndrome and Related Disorders," *Archives of Ophthalmology* 100, no. 2 (1982): 285–88.

14. Bruch, "Fröhlich Syndrome"; Solomon Solis-Cohen and Edward Weiss, "Dystrophia Adiposogenitalis, with Atypical Retinitis Pigmentosa and Mental Deficiency—the Laurence-Biedl Syndrome: A Report of Four Cases in One Family," *American Journal of the Medical Sciences* 169, no. 4 (1925): 499–504; E. A. Cockayne, D. Krestin, and A. Sorsby, "Obesity, Hypogenitalism, Mental Retardation, Polydactyly, and Retinal Pigmentation: The Laurence-Moon-Biedl Syndrome," *Quarterly Journal of Medicine* 4, no. 2 (1935): 93–120, 115.

15. R. L. Jenkins and H. G. Poncher, "Pathogenesis of the Laurence-Biedl Syndrome," *American Journal of Diseases of Children* 50, no. 1 (1935): 178–86.

16. A. Prader, A. Labhart, and H. Willi, "A Syndrome with Adiposity, Stunted Growth, Cryptocordia and Oligophrenia after Myotonia Entitled in Newborn," *Schweizerische Medizinische Wochenschrift* 86 (1956): 1260–61, English translation by Urs Eiholzer printed in appendix A of Butler, Lee, and Whitman, *Management of Prader-Willi Syndrome.*

17. "Dr. Andrea Prader," 2; Prader, Labhart, and Willi, "A Syndrome with Adiposity"; Prader et al., "Ein Sydrom von Adipositas."

18. A. Prader and H. Willi, "Das Syndrom von Imbezibilität, Adipositas, Muskelhypotonie, Hypogenitalismus, Hypogonadismus und Diabetes Mellitus mit 'myatonie'-Anamnese," *2nd International Congress of Mental Retardation, Vienna* (1961): 353–57.

19. B. M. Laurance, "Hypotonia, Obesity, Hypogonadism and Mental Retardation in Childhood," *Archives of Disease in Childhood* 36 (1961): 690; P. Royer, "Le diabète sucré

dans le syndrome de Willi-Prader," *Journee Annuelles de Diabetologie de l'Hotel-Dieu* 4 (1963): 91–99, 91.

20. P. R. Evans, "Hypogenital Dystrophy with Diabetic Tendency," *Guy's Hospital Reports* 113 (1964): 207–22, 221–22; H. Forssman and B. Hagberg, "Prader-Willi Syndrome in Boy of Ten with Prediabetes," *Acta Paediatrica* 53, no. 1 (1964): 70–78.

21. Zellweger and Schneider, "Syndrome of HHHO"; M. M. Cohen Jr. and R. J. Gorlin, "The Prader-Willi Syndrome," *Archives of Pediatrics and Adolescent Medicine* 117, no. 2 (1969): 213.

22. Cohen and Gorlin, "Prader-Willi Syndrome."

23. J. Roget, C. Mouriquand, Y. Bernard, J. Patet, J. Jobert, and C. Gilly, "Syndrome associant adiposité cryptorchidie et retard mental accompagné d'une aberration chromosomique," *Pediatrie* 20 (1965): 295–300; H. G. Dunn, D. K. Ford, N. Auersperg, and J. R. Miller, "Benign Congenital Hypotonia with Chromosomal Anomaly," *Pediatrics* 28, no. 4 (1961): 578–91.

24. Zellweger and Schneider, "Syndrome of HHHO."

25. C. J. Hawkey and A. Smithies, "The Prader-Willi Syndrome with a 15/15 Translocation: Case Report and Review of the Literature," *Journal of Medical Genetics* 13, no. 2 (1976): 152–63, 156.

26. J. J. Yunis, "High Resolution of Human Chromosomes," *Science* 191, no. 4233 (1976): 1268–70, 1270.

27. Ibid.; J. J. Yunis, J. R. Sawyer, and D. W. Ball, "The Characterization of High-Resolution G-Banded Chromosomes of Man," *Chromosoma* 67, no. 4 (1978): 293–307; D. E. Rooney and B. H. Czepulkowski, *Human Cytogenetics: A Practical Approach* (Oxford: IRL Press, 1986).

28. J. J. Yunis and M. E. Chandler, "High-Resolution Chromosome Analysis in Clinical Medicine," *Progress in Clinical Pathology* 7 (1977): 267–88; V. M. Riccardi, E. Sujansky, A. C. Smith, and U. Francke, "Chromosomal Imbalance in the Aniridia-Wilms' Tumor Association: 11p Interstitial Deletion," *Pediatrics* 61, no. 4 (1978): 604–10.

29. David H. Ledbetter, interview by author, Danville, PA, March 21, 2012; D. H. Ledbetter, "Discovery of the Chromosome 15 Deletion in Prader-Willi Syndrome," unpublished personal account, provided by David H. Ledbetter, July 2011, in Hogan's collection.

30. Ledbetter interview.

31. M. Fraccaro, O. Zuffardi, E. M. Buhler, and L. P. Jurik, "15/15 Translocation in Prader-Willi Syndrome," *Journal of Medical Genetics* 14, no. 4 (1977): 275–78, 275.

32. M. Kucerova, M. Strakova, and Z. Polivkova, "The Prader-Willi Syndrome with a 15/3 Translocation," *Journal of Medical Genetics* 16, no. 3 (1979): 234–35.

33. L. C. Strong, V. M. Riccardi, R. E. Ferrell, and R. S. Sparkes, "Familial Retinoblastoma and Chromosome 13 Deletion Transmitted via an Insertional Translocation," *Science* 213, no. 4515 (1981): 1501–3.

34. Ledbetter, "Discovery of the Chromosome."

35. Ibid., 6; F. C. Fraser, "Of Mice and Children: Reminiscences of a Teratogeneticist," *American Journal of Medical Genetics Part A* 146A, no. 17 (2008): 2179–202.

36. Ledbetter, "Discovery of the Chromosome."

37. D. H. Ledbetter, V. M. Riccardi, S. D. Airhart, R. J. Strobel, B. S. Keenan, and J. D. Crawford, "Deletions of Chromosome 15 as a Cause of the Prader–Willi Syndrome," *New England Journal of Medicine* 304, no. 6 (1981): 325–29.

38. Lorraine Daston and Peter Galison, *Objectivity* (New York: Zone Books, 2007).

39. Lorraine Daston and Peter Galison, "The Image of Objectivity," *Representations* 40, no. 3 (1992): 81–128; D. Warburton, D. A. Miller, O. J. Miller, P. W. Allderdice, and A. De Capoa, "Detection of Minute Deletions in Human Karyotypes," *Cytogenetic and Genome Research* 8, no. 2 (1969): 97–108. For more on standardization and subjectivity in human cytogenetics, see Susan Lindee, *Moments of Truth in Genetic Medicine* (Baltimore: Johns Hopkins University Press, 2005).

40. V. A. McKusick, "The Human Genome through the Eyes of Mercator and Vesalius," *Transactions of the American Clinical and Climatological Association* 92 (1981): 66–90, 79.

41. V. A. McKusick, "The Anatomy of the Human Genome," *Hospital Practice* 16, no. 4 (1981): 82–100. For more on the human genome as the "organ" of medical geneticists, see Andrew J. Hogan, "The 'Morbid Anatomy' of the Human Genome: Tracing the Observational and Representational Approaches of Postwar Genetics and Biomedicine," *Medical History* 58, no. 3 (2014): 315–36; Nathaniel Comfort, *The Science of Human Perfection: Heredity and Health in American Biomedicine* (New Haven, CT: Yale University Press, 2012).

42. Reed Pyeritz, interview by author, Philadelphia, April 18, 2012.

43. C. R. Scriver, "Window Panes of Eternity: Health, Disease, and Inherited Risk," *Yale Journal of Biology and Medicine* 55, nos. 5–6 (1982): 487–513, 497–98. On the history of genetic counseling, see Alexandra Minna Stern, *Telling Genes: The Story of Genetic Counseling in America* (Baltimore: Johns Hopkins University Press, 2012).

44. J. A. F. Roberts and M. E. Pembrey, *An Introduction to Medical Genetics*, 8th ed. (London: Oxford University Press, 1985), 235.

45. Gerald James Stine, *The New Human Genetics* (Dubuque, IA: Wm. C. Brown, 1989), 345–46.

46. Daniel J. Kevles, *In the Name of Eugenics: Genetics and the Uses of Human Heredity* (New York: Alfred A. Knopf, 1985), 249; L. S. Penrose, *Outline of Human Genetics*, 2nd ed. (New York: John Wiley, 1963), 136; Andrew J. Hogan, "Disrupting Genetic Dogma: Bridging Cytogenetics and Molecular Biology in Fragile X Research," *Historical Studies of the Natural Sciences* 45, no. 1 (2015): 174–97.

47. V. A. McKusick, "The Human Genome through the Eyes of a Clinical Geneticist," *Cytogenetic and Genome Research* 32, nos. 1–4 (1982): 7–23, 17.

48. Yunis, "High Resolution of Human Chromosomes."

49. D. H. Ledbetter and S. B. Cassidy, "The Etiology of Prader-Willi Syndrome: Clinical Implications of the Chromosome 15 Abnormalities," In *Prader-Willi Syndrome: Selected Research and Management Issues*, ed. Mary Lou Caldwell and Ronald L. Taylor (New York: Springer-Verlag, 1988), 13–28.

50. M. G. Butler, S. G. Kaler, P. Yu, and F. J. Meaney, "Metacarpophalangeal Pattern Profile Analysis in Prader-Willi Syndrome," *Clinical Genetics* 22, no. 6 (1982): 315–20.

51. J. F. Mattei, M. G. Mattei, and F. Giraud, "Prader-Willi Syndrome and Chromosome 15," *Human Genetics* 64, no. 4 (1983): 356–62; M. G. Mattei, N. Souiah, and J. F.

Mattei, "Chromosome 15 Anomalies and the Prader-Willi Syndrome: Cytogenetic Analysis," *Human Genetics* 66, no. 4 (1984): 313–34, 313.

52. H. F. de France, F. A. Beemer, and P. F. Ippel, "Duplication in Chromosome 15q in a Boy with the Prader-Willi Syndrome; Further Cytogenetic Confusion," *Clinical Genetics* 26, no. 4 (1984): 379–82; B. G. Kousseff and R. Douglass, "The Cytogenetic Controversy in the Prader-Labhart-Willi Syndrome," *American Journal of Medical Genetics* 13, no. 4 (1982): 431–39, 437.

53. D. H. Ledbetter, J. T. Mascarello, V. M. Riccardi, V. D. Harper, S. D. Airhart, and R. J. Strobel, "Chromosome 15 Abnormalities and the Prader-Willi Syndrome: A Follow-Up Report of 40 Cases," *American Journal of Human Genetics* 34, no. 2 (1982): 278–85.

54. V. A. McKusick, *Mendelian Inheritance in Man: Catalogs of Autosomal Dominant, Recessive, and X-Linked Phenotypes*, 2nd ed. (Baltimore: Johns Hopkins University Press, 1968), ix.

55. Maria Jesus Santesmases, "Size and the Centromere: Translocations and Visual Cultures in Early Human Genetics," in *Making Mutations: Objects, Practices, Contexts*, Preprint 393, ed. Luis Campos and Alexander von Schwerin (Berlin: Max Plank Institute for the History of Science, 2010), 189–208.

56. Ledbetter and Cassidy, "Etiology of Prader-Willi Syndrome."

57. Ibid.; A. Schinzel, "Approaches to the Prenatal Diagnosis of the Prader-Willi Syndrome." *Human Genetics* 74 (1986): 327; A. Smith, "Prenatal Diagnosis and the Prader-Willi Syndrome," *Human Genetics* 72, no. 3 (1986): 278.

58. F. Labidi and S. B. Cassidy, "A Blind Prometaphase Study of Prader-Willi Syndrome: Frequency and Consistency in Interpretation of Del 15q," *American Journal of Human Genetics* 39, no. 4 (1986): 452–60.

59. Ledbetter and Cassidy, "Etiology of Prader-Willi Syndrome"; Schinzel, "Approaches to the Prenatal Diagnosis."

60. S. B. Cassidy, H. C. Thuline, V. A. Holm, and J. M. Opitz, "Deletion of Chromosome 15 (Q11–13) in a Prader-Labhart-Willi Syndrome Clinic Population," *American Journal of Medical Genetics* 17, no. 2 (1984): 485–95.

61. R. D. Schmickel, "Contiguous Gene Syndromes: A Component of Recognizable Syndromes." *Journal of Pediatrics* 109, no. 2 (1986): 231–41; R. M. Pauli, L. F. Meisner, and R. J. Szmanda, "'Expanded' Prader-Willi Syndrome in a Boy with an Unusual 15q Chromosome Deletion," *American Journal of Diseases of Children* 137, no. 11 (1983): 1087–89.

62. Schmickel, "Contiguous Gene Syndromes," 236.

63. Ibid., 239.

64. McKusick, *Mendelian Inheritance in Man* (1968), ix.

65. Andrew J. Hogan, "Locating Genetic Disease: The Impact of Clinical Nosology on Biomedical Conceptions of the Human Genome (1966–1990)," *New Genetics and Society* 32, no. 1 (2013): 78–96.

66. R. E. Magenis, M. G. Brown, D. A. Lacy, S. Budden, S. LaFranchi, J. M. Opitz, J. F. Reynolds, and D. H. Ledbetter, "Is Angelman Syndrome an Alternate Result of Del (15) (q11q13)?," *American Journal of Medical Genetics* 28, no. 4 (1987): 829–38.

67. D. H. Ledbetter, F. Greenberg, V. A. Holm, S. B. Cassidy, J. M. Opitz, and J. F. Reynolds, "Second Annual Prader-Willi Syndrome Scientific Conference," *American Journal of Medical Genetics* 28, no. 4 (1987): 779–90.

68. Charles A. Williams, telephone interview by author, March 16, 2012.

69. Ledbetter et al., "Second Annual Prader-Willi Syndrome," 782.

70. L. C. Kaplan, R. Wharton, E. Elias, F. Mandell, T. Donlon, S. A. Latt, J. M. Opitz, and J. F. Reynolds, "Clinical Heterogeneity Associated with Deletions in the Long Arm of Chromosome 15: Report of 3 New Cases and Their Possible Genetic Significance," *American Journal of Medical Genetics* 28, no. 1 (1987): 45–53.

71. F. Greenberg, D. H. Ledbetter, J. M. Opitz, and J. F. Reynolds, "Deletions of Proximal 15q without Prader-Willi Syndrome," *American Journal of Medical Genetics* 28, no. 4 (1987): 813–20, 819.

72. Penrose, *The Biology of Mental Defect*, 136.

73. Ledbetter interview.

74. Stern, *Telling Genes;* Ilana Löwy, "How Genetics Came to the Unborn: 1960–2000," *Studies in History and Philosophy of Biological and Biomedical Sciences* 47 (2014): 290–99.

75. Ledbetter interview.

76. V. A. McKusick, *Mendelian Inheritance in Man: Catalogs of Autosomal Dominant, Autosomal Recessive, and X-Linked Phenotypes*, 5th ed. (Baltimore: Johns Hopkins University Press, 1978), xviii.

77. Daston and Galison, *Objectivity.*

Chapter 4. Seeing with Molecules

1. V. A. McKusick, "The Human Genome through the Eyes of Mercator and Vesalius," *Transactions of the American Clinical and Climatological Association* 92 (1981): 66–90, 76–77.

2. V. A. McKusick, "The Anatomy of the Human Genome," *American Journal of Medicine* 69, no. 2 (1980): 267–76, 272.

3. P. Berg, "Dissections and Reconstructions of Genes and Chromosomes," *Bioscience Reports* 1, no. 4 (1981): 269–87, 285.

4. Robert Nicholls, telephone interview by author, April 5, 2012; Andrew J. Hogan, "Disrupting Genetic Dogma: Bridging Cytogenetics and Molecular Biology in Fragile X Research," *Historical Studies of the Natural Sciences* 45, no. 1 (2015): 174–97.

5. Hans-Jorg Rheinberger, *Toward a History of Epistemic Things: Synthesizing Proteins in the Test Tube* (Stanford, CA: Stanford University Press, 1997).

6. Peter Keating and Alberto Cambrosio, "Does Biomedicine Entail the Successful Reduction of Pathology to Biology?," *Perspectives in Biology and Medicine* 47, no. 3 (2004): 357–71; Peter Keating and Alberto Cambrosio, *Biomedical Platforms: Realigning the Normal and the Pathological in Late-Twentieth Century Medicine* (Cambridge, MA: MIT Press, 2003); Peter Keating and Alberto Cambrosio, "Signs, Markers, Profiles, and Signatures: Clinical Haemotology Meets the New Genetics (1980–2000)," *New Genetics and Society* 23, no. 1 (2004): 15–45, 26; Soraya de Chadarevian, *Designs for Life: Molecular Biology after World War II* (Cambridge: Cambridge University Press, 2002); Soraya de Chadarevian and Harmke Kamminga, eds., *Molecularizing Biology and Medicine: New Prac-*

tices and Alliances, 1910s–1970s (Amsterdam: Harwood Academic, 1998); Ilana Löwy, *Between Bench and Bedside: Science, Healing, and Interleukin-2 in a Cancer Ward* (Cambridge, MA: Harvard University Press, 1996).

7. J. D. Rowley, "Molecular Cytogenetics: Rosetta Stone for Understanding Cancer—Twenty-Ninth G. H. A. Clowes Memorial Award Lecture," *Cancer Research* 50, no. 13 (1990): 3816–25.

8. J. J. Yunis, "High Resolution of Human Chromosomes," *Science* 191, no. 4233 (1976): 1268–70, 1270.

9. Bruno J. Strasser, "A World in One Dimension: Linus Pauling, Francis Crick and the Central Dogma of Molecular Biology," *History and Philosophy of the Life Sciences* 28, no. 4 (2006): 491–512; Uta Francke, interview by author, Palo Alto, CA, February 27, 2012.

10. R. J. Britten and D. E. Kohne, "Repeated Sequences in DNA," *Science* 161, no. 3841 (1968): 529–40; C. L. Schildkraut, J. Marmur, and P. Doty, "The Formation of Hybrid DNA Molecules and Their Use in Studies of DNA Homologies," *Journal of Molecular Biology* 3, no. 5 (1961): 595–617.

11. As historian Angela Creager has insightfully demonstrated, radiolabeled tracers were widely adopted in postwar research, as physicists turned to biology and as the Atomic Energy Commission sought new uses for isotopes, with the political aim of transforming nuclear physics from a science of death into a science of life. Angela N. H. Creager, *Life Atomic: A History of Radioisotopes in Science and Medicine* (Chicago: University of Chicago Press, 2013).

12. M. L. Pardue and J. G. Gall, "Molecular Hybridization of Radioactive DNA to the DNA of Cytological Preparations," *Proceedings of the National Academy of Sciences* 64, no. 2 (1969): 600–604; M. L. Pardue and J. G. Gall, "Chromosomal Localization of Mouse Satellite DNA," *Science* 168, no. 3937 (1970): 1356–58.

13. J. G. Gall and M. L. Pardue, "Formation and Detection of RNA-DNA Hybrid Molecules in Cytological Preparations," *Proceedings of the National Academy of Sciences* 63, no. 2 (1969): 378–83.

14. P. M. Price, J. H. Conover, and K. Hirschhorn, "Chromosomal Localization of Human Haemoglobin Structural Genes," *Nature* 237, no. 5354 (1972): 340–42, 340.

15. P. M. Price and K. Hirschhorn, "In situ Hybridization for Gene Mapping," *Cytogenetics and Cell Genetics* 14, nos. 3–6 (1975): 395–401, 400.

16. The years between 1975 and 1985 saw the widespread adoption of prenatal diagnosis and the development of many new techniques, which were of variable safety and efficacy. Andrew J. Hogan, "Set Adrift in the Prenatal Diagnostic Marketplace: Analyzing the Role of Users and Mediators in the History of a Medical Technology," *Technology and Culture* 54, no. 1 (2013): 62–89; Ruth Schwartz Cowan, "Women's Roles in the History of Amniocentesis and Chorionic Villus Sampling," in *Women and Prenatal Testing: Facing the Challenges of Genetic Technology*, ed. Karen H. Rothenberg and Elizabeth Jean Thomson (Columbus: Ohio State University Press, 1994), 35–48; B. P. S. Friedman Alter, J. C. Hobbins, M. J. Mahoney, A. S. Sherman, J. F. McSweeney, D. G. Nathan, and E. Schwartz, "Prenatal Diagnosis of Sickle-Cell Anemia and Alpha G Philadelphia: Study of a Fetus Also at Risk for Hb S/beta+–Thalassemia," *New England Journal of Medicine* 294, no. 19 (1976): 1040–41.

17. Y. W. Kan, M. S. Golbus, and A. M. Dozy, "Prenatal Diagnosis of Alpha-Thalassemia: Clinical Application of Molecular Hybridization," *New England Journal of Medicine* 295, no. 21 (1976): 1165–67.

18. D. G. Nathan, "Antenatal Diagnosis of Hemoglobinopathies: An Exquisite Molecular Brew," *New England Journal of Medicine* 295, no. 21 (1976): 1196–98.

19. The timing of an abortion had an impact on the procedure used, its invasiveness and risk, and the potential social challenges and stigma associated with late-term abortion. Rayna Rapp, *Testing Women, Testing the Fetus* (New York: Routledge, 1999). As I show elsewhere, physicians at this time overestimated the demand for earlier prenatal diagnostic options. Hogan, "Set Adrift."

20. McKusick, "Human Genome through the Eyes of Mercator," 76–77; V. A. McKusick, "Diseases of the Genome," *Journal of the American Medical Association* 252, no. 8 (1984): 1041–48; V. A. McKusick, "The Human Genome through the Eyes of a Clinical Geneticist," *Cytogenetic and Genome Research* 32, nos. 1–4 (1982): 7–23.

21. T. J. Kelly and H. O. Smith, "A Restriction Enzyme for *Hemophilus Influenzae:* II, Base Sequence of the Recognition Site," *Journal of Molecular Biology* 51, no. 2 (1970): 393–409; H. O. Smith and K. W. Wilcox, "A Restriction Enzyme from Hemophilus Influenzae: I, Purification and General Properties," *Journal of Molecular Biology* 51, no. 2 (1970): 379–91; K. Danna and D. Nathans, "Specific Cleavage of Simian Virus 40 DNA by Restriction Endonuclease of Hemophilus Influenzae," *Proceedings of the National Academy of Sciences* 68, no. 12 (1971): 2913–17.

22. S. N. Cohen, A. C. Chang, H. W. Boyer, and R. B. Helling, "Construction of Biologically Functional Bacterial Plasmids in vitro," *Proceedings of the National Academy of Sciences* 70, no. 11 (1973): 3240–44, 3244; Doogab Yi, *The Recombinant University: Genetic Engineering and the Emergence of Stanford Biotechnology* (Chicago: University of Chicago Press, 2015).

23. Y. W. Kan and A. M. Dozy, "Polymorphism of DNA Sequence Adjacent to Human Beta-Globin Structural Gene: Relationship to Sickle Mutation," *Proceedings of the National Academy of Sciences* 75, no. 11 (1978): 5631–35, 5631.

24. Y. W. Kan and A. M. Dozy, "Antenatal Diagnosis of Sickle-Cell Anaemia by DNA Analysis of Amniotic-Fluid Cells," *Lancet* 2, no. 8096 (1978): 910–12.

25. Georges Canguillhem, *The Normal and the Pathological* (New York: Zone, 1991).

26. Britten and Kohne, "Repeated Sequences in DNA"; S. Ohno, "So Much 'Junk' DNA in Our Genome," *Brookhaven Symposiums in Biology* 23 (1972): 366–70.

27. R. M. Lawn, E. F. Fritsch, R. C. Parker, G. Blake, and T. Maniatis, "The Isolation and Characterization of Linked Delta- and Beta-Globin Genes from a Cloned Library of Human DNA," *Cell* 15, no. 4 (1978): 1157–74; T. Maniatis, R. C. Hardison, E. Lacy, J. Lauer, C. O'Connell, D. Quon, G. K. Sim, and A. Efstratiadis, "The Isolation of Structural Genes from Libraries of Eucaryotic DNA," *Cell* 15, no. 2 (1978): 687–701.

28. W. D. Benton and R. W. Davis, "Screening Lambda-Gt Recombinant Clones by Hybridization to Single Plaques in situ," *Science* 196, no. 4286 (1977): 180–82.

29. A. V. Carrano, J. W. Gray, R. G. Langlois, K. J. Burkhart-Schultz, and M. A. Van Dilla, "Measurement and Purification of Human Chromosomes by Flow Cytometry and Sorting," *Proceedings of the National Academy of Sciences* 76, no. 3 (1979): 1382–84; J. W.

Gray, A. V. Carrano, L. L. Steinmetz, M. A. Van Dilla, D. H. Moore 2nd, B. H. Mayall, and M. L. Mendelsohn, "Chromosome Measurement and Sorting by Flow Systems," *Proceedings of the National Academy of Sciences* 72, no. 4 (1975): 1231–34.

30. K. E. Davies, B. D. Young, R. G. Elles, M. E. Hill, and R. Williamson, "Cloning of a Representative Genomic Library of the Human X Chromosome after Sorting by Flow Cytometry," *Nature* 293, no. 5831 (1981): 374–76.

31. D. Botstein, R. L. White, M. Skolnick, and R. W. Davis, "Construction of a Genetic Linkage Map in Man Using Restriction Fragment Length Polymorphisms," *American Journal of Human Genetics* 32, no. 3 (1980): 314–31. Peter Keating and Alberto Cambrosio pointed to this work as representative of the often overlooked contributions of clinical research in molecular biology. Kan and Dozy, they argued, were not simply applying tools from molecular biology in the clinic. Rather, their clinical work directly contributed to basic advances in the field. Peter Keating and Alberto Cambrosio, "The New Genetics and Cancer: The Contributions of Clinical Medicine in the Era of Biomedicine," *Journal of the History of Medicine and Allied Sciences* 56, no. 4 (2001): 321–52.

32. Botstein et al., "Construction of a Genetic Linkage Map."

33. Keating and Cambrosio, "New Genetics and Cancer"; L. C. Dunn, "Old and New in Genetics," *Bulletin of the New York Academy of Medicine* 40, no. 5 (1964): 325–33.

34. D. E. Comings, "Prenatal Diagnosis and the 'New Genetics,'" *American Journal of Human Genetics* 32, no. 3 (1980): 453–54, 453.

35. McKusick, "Anatomy of the Human Genome," *American Journal of Medicine,* 276.

36. N. Wexler, "Clairvoyance and Caution: Repercussions from the Human Genome Project," in *Code of Codes: Scientific and Social Issues in the Human Genome Project,* ed. Daniel J. Kevles and Leroy Hood (Cambridge, MA: Harvard University Press, 1992), 211–43, 221.

37. J. F. Gusella, N. S. Wexler, P. M. Conneally, S. L. Naylor, M. A. Anderson, R. E. Tanzi, P. C. Watkins, K. Ottina, M. R. Wallace, A. Y. Sakaguchi, et al., "A Polymorphic DNA Marker Genetically Linked to Huntington's Disease," *Nature* 306, no. 5940 (1983): 234–38.

38. Ibid.

39. Ibid., 238.

40. R. G. Knowlton, O. Cohen-Haguenauer, N. Van Cong, J. Frezal, V. A. Brown, D. Barker, J. C. Braman, J. W. Schumm, L. C. Tsui, M. Buchwald, et al., "A Polymorphic DNA Marker Linked to Cystic Fibrosis Is Located on Chromosome 7," *Nature* 318, no. 6044 (1985): 380–82; L. C. Tsui, M. Buchwald, D. Barker, J. C. Braman, R. Knowlton, J. W. Schumm, H. Eiberg, J. Mohr, D. Kennedy, N. Plavsic, et al., "Cystic Fibrosis Locus Defined by a Genetically Linked Polymorphic DNA Marker," *Science* 230, no. 4729 (1985): 1054–57; B. J. Wainwright, P. J. Scambler, J. Schmidtke, E. A. Watson, H. Y. Law, M. Farrall, H. J. Cooke, H. Eiberg, and R. Williamson, "Localization of Cystic Fibrosis Locus to Human Chromosome 7cen-Q22," *Nature* 318, no. 6044 (1985): 384–85; R. White, S. Woodward, M. Leppert, P. O'Connell, M. Hoff, J. Herbst, J. M. Lalouel, M. Dean, and G. Vande Woude, "A Closely Linked Genetic Marker for Cystic Fibrosis," *Nature* 318, no. 6044 (1985): 382–84.

41. J. J. Holden, H. S. Wang, and B. N. White, "The Fragile-X Syndrome: IV, Progress towards the Identification of Linked Restriction Fragment Length Variants (RFLVs)," *American Journal of Medical Genetics* 17, no. 1 (1984): 259–73; L. R. Shapiro, P. L. Wilmot, P. D. Murphy, and W. R. Breg, "Experience with Multiple Approaches to the Prenatal Diagnosis of the Fragile X Syndrome: Amniotic Fluid, Chorionic Villi, Fetal Blood and Molecular Methods," *American Journal of Medical Genetics* 30, nos. 1–2 (1988): 347–54. For more see Andrew J. Hogan, "Visualizing Carrier Status: Fragile X Syndrome and Genetic Diagnosis since the 1940s," *Endeavour* 36, no. 2 (2012): 77–84; Hogan, "Disrupting Genetic Dogma."

42. David H. Ledbetter, interview by author, Danville, PA, March 21, 2012.

43. T. A. Donlon, M. Lalande, A. Wyman, G. Bruns, and S. A. Latt, "Isolation of Molecular Probes Associated with the Chromosome 15 Instability in the Prader-Willi Syndrome," *Proceedings of the National Academy of Sciences* 83, no. 12 (1986): 4408–12.

44. U. Tantravahi, R. D. Nicholls, H. Stroh, S. Ringer, R. L. Neve, L. Kaplan, R. Wharton, D. Wurster-Hill, J. M. Graham Jr., E. S. Cantu, et al., "Quantitative Calibration and Use of DNA Probes for Investigating Chromosome Abnormalities in the Prader-Willi Syndrome," *American Journal of Medical Genetics* 33, no. 1 (1989): 78–87.

45. R. D. Nicholls, J. H. M. Knoll, M. G. Butler, S. Karam, and M. Lalande, "Genetic Imprinting Suggested by Maternal Heterodisomy in Non-Deletion Prader-Willi Syndrome," *Nature* 342, no. 6247 (1989): 281–85; J. H. M. Knoll, R. D. Nicholls, R. E. Magenis, J. M. Graham Jr, M. Lalande, S. A. Latt, J. M. Opitz, and J. F. Reynolds, "Angelman and Prader-Willi Syndromes Share a Common Chromosome 15 Deletion but Differ in Parental Origin of the Deletion," *American Journal of Medical Genetics* 32, no. 2 (1989): 285–90; R. D. Nicholls, J. H. Knoll, K. Glatt, J. H. Hersh, T. D. Brewster, J. M. Graham Jr., D. Wurster-Hill, R. Wharton, and S. A. Latt, "Restriction Fragment Length Polymorphisms within Proximal 15q and Their Use in Molecular Cytogenetics and the Prader-Willi Syndrome," *American Journal of Medical Genetics* 33, no. 1 (1989): 66–77.

46. R. D. Nicholls, "Genomic Imprinting and Uniparental Disomy in Angelman and Prader-Willi Syndromes: A Review," *American Journal of Medical Genetics* 46, no. 1 (1993): 16–25.

47. M. G. Butler and C. G. Palmer, "Parental Origin of Chromosome 15 Deletion in Prader-Willi Syndrome," *Lancet* 1, no. 8336 (1983): 1285–86; R. E. Magenis, M. G. Brown, D. A. Lacy, S. Budden, S. LaFranchi, J. M. Opitz, J. F. Reynolds, and D. H. Ledbetter, "Is Angelman Syndrome an Alternate Result of Del (15)(qllql3)?," *American Journal of Medical Genetics* 28, no. 4 (1987): 829–38.

48. Ledbetter interview.

49. For more on this, see Andrew J. Hogan, "Locating Genetic Disease: The Impact of Clinical Nosology on Biomedical Conceptions of the Human Genome (1966–1990)," *New Genetics and Society* 32, no. 1 (2013): 78–96.

50. Creager, *Life Atomic;* J. G. Bauman, J. Wiegant, P. Van Duijn, N. H. Lubsen, P. J. Sondermeijer, W. Hennig, and E. Kubli, "Rapid and High Resolution Detection of in situ Hybridisation to Polytene Chromosomes Using Fluorochrome-Labeled RNA," *Chromosoma* 84, no. 1 (1981): 1–18.

51. Rowley, "Molecular Cytogenetics"

52. M. E. Harper and G. F. Saunders, "Localization of Single Copy DNA Sequences of G-Banded Human Chromosomes by in situ Hybridization," *Chromosoma* 83, no. 3 (1981): 431–39; M. E. Harper, A. Ullrich, and G. F. Saunders, "Localization of the Human Insulin Gene to the Distal End of the Short Arm of Chromosome 11," *Proceedings of the National Academy of Sciences* 78, no. 7 (1981): 4458–60.

53. C. R. Alfageme, G. T. Rudkin, and L. H. Cohen, "Locations of Chromosomal Proteins in Polytene Chromosomes," *Proceedings of the National Academy of Sciences* 73, no. 6 (1976): 2038–42, 2039.

54. G. T. Rudkin and B. D. Stollar, "High Resolution Detection of DNA-RNA Hybrids in situ by Indirect Immunofluorescence," *Nature* 265, no. 5593 (1977): 472–73, 473.

55. S. W. Cheung, P. V. Tishler, L. Atkins, S. K. Sengupta, E. J. Modest, and B. G. Forget, "Gene Mapping by Fluorescent in situ Hybridization," *Cell Biology International Reports* 1, no. 3 (1977): 255–62; J. G. Bauman, J. Wiegant, P. Borst, and P. van Duijn, "A New Method for Fluorescence Microscopical Localization of Specific DNA Sequences by in situ Hybridization of Fluorochromelabelled RNA," *Experimental Cell Research* 128, no. 2 (1980): 485–90.

56. T. R. Broker, L. M. Angerer, P. H. Yen, N. D. Hershey, and N. Davidson, "Electron Microscopic Visualization of tRNA Genes with Ferritin-Avidin: Biotin Labels," *Nucleic Acids Research* 5, no. 2 (1978): 363–84; M. Wu and N. Davidson, "Transmission Electron Microscopic Method for Gene Mapping on Polytene Chromosomes by in situ Hybridization," *Proceedings of the National Academy of Sciences* 78, no. 11 (1981): 7059–63.

57. P. R. Langer-Safer, M. Levine, and D. C. Ward, "Immunological Method for Mapping Genes on Drosophila Polytene Chromosomes," *Proceedings of the National Academy of Sciences* 79, no. 14 (1982): 4381–85.

58. Ibid.

59. J. E. Landegent, N. Jansen in de Wal, Y. M. Fisser-Groen, E. Bakker, M. van der Ploeg, and P. L. Pearson, "Fine Mapping of the Huntington Disease Linked D4s10 Locus by Non-Radioactive in situ Hybridization," *Human Genetics* 73, no. 4 (1986): 354–57; J. E. Landegent, N. Jansen in de Wal, G. J. van Ommen, F. Baas, J. J. de Vijlder, P. van Duijn, and M. Van der Ploeg, "Chromosomal Localization of a Unique Gene by Non-Autoradiographic in situ Hybridization," *Nature* 317, no. 6033 (1985): 175–77.

60. R. E. Magenis, J. Gusella, K. Weliky, S. Olson, G. Haight, S. Toth-Fejel, and R. Sheehy, "Huntington Disease-Linked Restriction Fragment Length Polymorphism Localized within Band P16.1 of Chromosome 4 by in situ Hybridization," *American Journal of Human Genetics* 39, no. 3 (1986): 383–91.

61. D. Pinkel, J. Landegent, C. Collins, J. Fuscoe, R. Segraves, J. Lucas, and J. Gray, "Fluorescence in situ Hybridization with Human Chromosome-Specific Libraries: Detection of Trisomy 21 and Translocations of Chromosome 4," *Proceedings of the National Academy of Sciences* 85, no. 23 (1988): 9138–42; D. Pinkel, T. Straume, and J. W. Gray, "Cytogenetic Analysis Using Quantitative, High-Sensitivity, Fluorescence Hybridization," *Proceedings of the National Academy of Sciences* 83, no. 9 (1986): 2934–38.

62. Yunis, "High Resolution of Human Chromosomes"; B. J. Trask, "Fluorescence in situ Hybridization: Applications in Cytogenetics and Gene Mapping," *Trends in Genetics* 7, no. 5 (1991): 149–54.

63. Rowley, "Molecular Cytogenetics."

64. Michel Morange, *A History of Molecular Biology* (Cambridge, MA: Harvard University Press, 1998), 244–45.

65. For other examples of molecularization during the twentieth century, see de Chadarevian and Kamminga, *Molecularizing Biology and Medicine;* Hogan, "Disrupting Genetic Dogma."

66. Horace F. Judson, "A History of Gene Mapping and Sequencing," in Kevles and Hood, *Code of Codes,* 72.

Chapter 5. Institutionalized Disorders

1. V. A. McKusick, "On Lumpers and Splitters, or the Nosology of Genetic Disease," *Birth Defects* 5, no. 1 (1969): 23–32.

2. Ibid., 23–24. Splitting, however, continued to be important in medical genetics, and it was increasingly considered as clinically similar disorders were found to have distinct genetic causes.

3. R. E. Pyeritz, "Marfan Syndrome: 30 Years of Research Equals 30 Years of Additional Life Expectancy," *Heart* 95, no. 3 (2009): 173–75.

4. Paul Rabinow, "Artificiality and Enlightenment: From Sociobiology to Biosociality," in *Anthropologies of Modernity: Foucault, Governmentality, and Life Politics,* ed. J. X. Inda (Oxford: Blackwell, 1992), 179–93; Daniel Navon and U. Shwed, "The Chromosome 22q11.2 Deletion: From the Unification of Biomedical Fields to a New Kind of Genetic Condition," *Social Science & Medicine* 75, no. 9 (2012): 1633–41; Vololona Rabeharisoa, Michel Callon, Angela Marques Filipe, Joao Arriscado Nunes, Florence Paterson, and Frederic Vergnaud, "From 'Politics of Numbers' to 'Politics of Singularization': Patients' Activism and Engagement in Research on Rare Diseases in France and Portugal," *BioSocieties* 9, no. 2 (2014): 194–217.

5. V. A. McKusick, "The Human Genome through the Eyes of a Clinical Geneticist," *Cytogenetic and Genome Research* 32, nos. 1–4 (1982): 7–23, 20.

6. A. M. DiGeorge, "Discussion of Cooper *et al.*, 'A New Concept of the Cellular Basis of Immunity,'" *Journal of Pediatrics* 67, no. 5 (1965): 907–8; A. M. DiGeorge, "Congenital Absence of the Thymus and Its Immunologic Consequences: Concurrence with Congenital Hypoparathyroidism," *Birth Defects Original Article Series* 4, no. 1 (1968): 116–23.

7. DiGeorge, "Congenital Absence of the Thymus," 117.

8. Ibid., 120.

9. For more on the history of bioethical oversight in the clinical setting, see Albert R. Jonsen, *The Birth of Bioethics* (New York: Oxford University Press, 2003); Laura Stark, *Behind Closed Doors: IRBs and the Making of Ethical Research* (Chicago: University of Chicago Press, 2011); David J. Rothman, *Strangers at the Bedside: A History of How Law and Bioethics Transformed Medical Decision Making* (New York: Basic Books, 1991).

10. Steven J. Peitzman, "From Dropsy to Bright's Disease to End-Stage Renal Disease," *Milbank Quarterly* 67 (1989): 16–32; Andrew J. Hogan, "Medical Eponyms: Patient Advocates, Professional Interests, and the Persistence of Honorary Naming," *Social History of Medicine,* forthcoming in 2016.

11. DiGeorge, "Congenital Absence of the Thymus," 120; E. J. Lammer, J. M. Opitz, and J. F. Reynolds, "The DiGeorge Anomaly as a Developmental Field Defect," *American Journal of Medical Genetics* 25, no. S2 (1986): 113–27, 114.

12. W. E. Dodson, D. Alexander, M. Al-Aish, and F. De La Cruz, "The DiGeorge Syndrome," *Lancet* 1, no. 7594 (1969): 574–75; R. M. Freedom, F. S. Rosen, and A. S. Nadas, "Congenital Cardiovascular Disease and Anomalies of the Third and Fourth Pharyngeal Pouch," *Circulation* 46, no. 1 (1972): 165–72; J. C. Harvey, W. T. Dungan, M. J. Elders, and E. R. Hughes, "Third and Fourth Pharyngeal Pouch Syndrome, Associated Vascular Anomalies and Hypocalcemic Seizures," *Clinical Pediatrics* 9, no. 8 (1970): 496–99; R. Kretschmer, B. Say, D. Brown, and F. S. Rosen, "Congenital Aplasia of the Thymus Gland (DiGeorge's Syndrome)," *New England Journal of Medicine* 279, no. 24 (1968): 1295–1301.

13. C. S. August, A. I. Berkel, R. H. Levey, F. S. Rosen, and H. E. Kay, "Establishment of Immunological Competence in a Child with Congenital Thymic Aplasia by a Graft of Fetal Thymus," *Lancet* 1, no. 7656 (1970): 1080–83; C. S. August, F. S. Rosen, R. M. Filler, C. A. Janeway, B. Markowski, and H. E. Kay, "Implantation of a Foetal Thymus, Restoring Immunological Competence in a Patient with Thymic Aplasia (DiGeorge's Syndrome)," *Lancet* 2, no. 7580 (1968): 1210–11; W. W. Cleveland, B. J. Fogel, W. T. Brown, and H. E. M. Kay, "Foetal Thymic Transplant in a Case of DiGeorge's Syndrome," *Lancet* 292, no. 7580 (1968): 1211–14.

14. H. W. Lischner, "DiGeorge Syndrome(s)," *Journal of Pediatrics* 81, no. 5 (1972): 1042.

15. DiGeorge, "Congenital Absence of the Thymus," 116.

16. J. P. Finley, G. F. Collins, J. P. De Chadarevian, and R. L. Williams, "DiGeorge Syndrome Presenting as Severe Congenital Heart Disease in the Newborn," *Canadian Medical Association Journal* 116, no. 6 (1977): 635–40, 637.

17. M. E. Conley, J. B. Beckwith, J. F. K. Mancer, and L. Tenckhoff, "The Spectrum of the DiGeorge Syndrome," *Journal of Pediatrics* 94, no. 6 (1979): 883–90.

18. Ibid.; Pyeritz, "Marfan Syndrome."

19. J. C. Carey, "Spectrum of DiGeorge 'Syndrome,'" *Journal of Pediatrics* 96, no. 5 (1980): 955–56, 955.

20. Ibid., 955; Kretschmner et al., "Congenital Aplasia of the Thymus."

21. Kurt Hirschhorn, interview by author, New York, January 26, 2012.

22. McKusick, "On Lumpers and Splitters."

23. J. M. Opitz, J. Herrmann, J. C. Pettersen, E. T. Bersu, and S. C. Colacino, "Terminological, Diagnostic, Nosological, and Anatomical-Developmental Aspects of Developmental Defects in Man," *Advances in Human Genetics* 9 (1979): 71–164, 95; D. W. Smith, *Recognizable Patterns of Human Malformation* (Philadelphia: W. B. Saunders, 1970); Carey, "Spectrum of DiGeorge 'Syndrome'"; D. W. Smith, "Dysmorphology (Teratology)," *Journal of Pediatrics* 69, no. 6 (1966): 1150–69, 1151.

24. Nathaniel Comfort, *The Science of Human Perfection: Heredity and Health in American Biomedicine* (New Haven, CT: Yale University Press, 2012), 91.

25. McKusick, "Human Genome through the Eyes of a Geneticist," 20.

26. DiGeorge, "Congenital Absence of the Thymus." Peter Nowell and David Hungerford at the University of Pennsylvania were also conducting clinically oriented human chromosomal analysis at this time. In 1960 they reported on a chromosomal abnormality

that came to be known as the Philadelphia chromosome and on its role in causing chronic myeloid leukemia. P. C. Nowell and D. A. Hungerford, "Chromosome Studies on Normal and Leukemic Human Leukocytes," *Journal of the National Cancer Institute* 25, no. 1 (1960): 85–109.

27. R. W. Steele, C. Limas, G. B. Thurman, M. Schuelein, H. Bauer, and J. A. Bellanti, "Familial Thymic Aplasia," *New England Journal of Medicine* 287, no. 16 (1972): 787–91.

28. M. Raatikka, J. Rapola, L. Tuuteri, I. Louhimo, and E. Savilahti, "Familial Third and Fourth Pharyngeal Pouch Syndrome with Truncus Arteriosus: DiGeorge Syndrome," *Pediatrics* 67, no. 2 (1981): 173–75.

29. A. de la Chapelle, R. Herva, M. Koivisto, and P. Aula, "A Deletion in Chromosome 22 Can Cause DiGeorge Syndrome," *Human Genetics* 57, no. 3 (1981): 253–56, 254–55.

30. Ibid., 255.

31. Albert de la Chapelle, telephone interview by author, March 25, 2015.

32. V. M. Riccardi, E. Sujansky, A. C. Smith, and U. Francke, "Chromosomal Imbalance in the Aniridia-Wilms' Tumor Association: 11p Interstitial Deletion," *Pediatrics* 61, no. 4 (1978): 604–10; D. H. Ledbetter, V. M. Riccardi, S. D. Airhart, R. J. Strobel, B. S. Keenan, and J. D. Crawford, "Deletions of Chromosome 15 as a Cause of the Prader-Willi Syndrome," *New England Journal of Medicine* 304, no. 6 (1981): 325–29; McKusick, "Human Genome through the Eyes of a Geneticist," 20.

33. McKusick, "Human Genome through the Eyes of a Geneticist," 20.

34. Beverly Emanuel, interview by author, Philadelphia, November 9, 2011; Elaine H. Zackai, Beverly S. Emanuel, and John M. Optiz, "Site-Specific Reciprocal Translocation, t(11; 22)(q23; Q11), in Several Unrelated Families with 3:1 Meiotic Disjunction," *American Journal of Medical Genetics* 7, no. 4 (1980): 507–21; Hogan, "Medical Eponyms."

35. Emanuel interview.

36. R. I. Kelley, E. H. Zackai, B. S. Emanuel, M. Kistenmacher, F. Greenberg, and H. H. Punnett, "The Association of the DiGeorge Anomalad with Partial Monosomy of Chromosome 22+," *Journal of Pediatrics* 101, no. 2 (1982): 197–200.

37. F. Greenberg, W. E. Crowder, V. Paschall, J. Colon-Linares, B. Lubianski, and D. H. Ledbetter, "Familial DiGeorge Syndrome and Associated Partial Monosomy of Chromosome 22," *Human Genetics* 65, no. 4 (1984): 317–19, 318.

38. Emanuel interview.

39. For more on the interest of molecular biologists in studying these disorders, see Andrew J. Hogan, "Locating Genetic Disease: The Impact of Clinical Nosology on Biomedical Conceptions of the Human Genome (1966–1990)," *New Genetics and Society* 32, no. 1 (2013): 78–96; Andrew J. Hogan, "Disrupting Genetic Dogma: Bridging Cytogenetics and Molecular Biology in Fragile X Research," *Historical Studies of the Natural Sciences* 45, no. 1 (2015): 174–97.

40. Robert Shprintzen, telephone interview by author, March 26, 2015. On the history of genetic counseling, see Alexandra Minna Stern, *Telling Genes: The Story of Genetic Counseling in America* (Baltimore: Johns Hopkins University Press, 2012).

41. Shprintzen interview.

42. R. J. Shprintzen, R. B. Goldberg, M. L. Lewin, E. J. Sidoti, M. D. Berkman, R. V. Argamaso, and D. Young, "A New Syndrome Involving Cleft Palate, Cardiac Anomalies,

Typical Facies, and Learning Disabilities: Velo-Cardio-Facial Syndrome," *Cleft Palate Journal* 5, no. 1 (1978): 56–62, 57. The observation by clinicians that children with the same disorder looked more like one another than like their individual parents was common in clinical delineation, going back to John Langdon Down in the 1860s, who described what was later called Down syndrome. Parents often refuted this characterization. P. M. Dunn, "Dr. Langdon Down (1828–1896) and 'Mongolism,'" *Archives of Disease in Childhood* 66, no. 7 (1991): 827–28; Hogan, "Medical Eponyms."

43. Shprintzen interview.

44. Shprintzen et al., "New Syndrome Involving Cleft Palate."

45. Smith, "Dysmorphology"; Shprintzen et al., "New Syndrome Involving Cleft Palate," 60.

46. R. J. Shprintzen, R. B. Goldberg, D. Young, and L. Wolford, "The Velo-Cardio-Facial Syndrome: A Clinical and Genetic Analysis," *Pediatrics* 67, no. 2 (1981): 167–72; D. W. Smith and K. L. Jones, *Recognizable Patterns of Human Malformation: Genetic, Embryologic, and Clinical Aspects,* 3rd ed. (Philadelphia: W. B. Saunders, 1982), 194.

47. R. Goldberg, R. Marion, M. Borderon, A. Wiznia, and R. J. Shprintzen, "Phenotypic Overlap between Velo-Cardio-Facial Syndrome and the DiGeorge Sequence," *American Journal of Human Genetics* 37 (1985): A54.

48. Lammer, Opitz, and Reynolds, "DiGeorge Anomaly," 115.

49. Opitz et al., "Terminological, Diagnostic, Nosological," 98.

50. Carey, "Spectrum of DiGeorge 'Syndrome'"; C. A. Stevens, J. C. Carey, and A. O. Shigeoka, "DiGeorge Anomaly and Velocardiofacial Syndrome," *Pediatrics* 85, no. 4 (1990): 526–30, 526.

51. Carey, "Spectrum of the DiGeorge 'Syndrome'," 955.

52. F. Greenberg, F. F. Elder, P. Haffner, H. Northrup, and D. H. Ledbetter, "Cytogenetic Findings in a Prospective Series of Patients with DiGeorge Anomaly," *American Journal of Human Genetics* 43, no. 5 (1988): 605–11; J. T. Mascarello, J. F. Bastian, M. C. Jones, J. M. Opitz, and J. F. Reynolds, "Interstitial Deletion of Chromosome 22 in a Patient with the DiGeorge Malformation Sequence," *American Journal of Medical Genetics* 32, no. 1 (1989): 112–14; A. Lipson, K. Fagan, A. Colley, P. Colley, G. Sholler, D. Issacs, and R. K. Oates, "Velo-Cardio-Facial and Partial DiGeorge Phenotype in a Child with Interstitial Deletion at 10p13—implications for Cytogenetics and Molecular Biology," *American Journal of Medical Genetics* 65, no. 4 (1996): 304–8; S. C. Daw, C. Taylor, M. Kraman, K. Call, J. Mao, S. Schuffenhauer, T. Meitinger, T. Lipson, J. Goodship, and P. Scambler, "A Common Region of 10p Deleted in DiGeorge and Velocardiofacial Syndromes," *Nature Genetics* 13, no. 4 (1996): 458–60. For more on genetic heterogeneity in postwar medical genetics, see McKusick, "On Lumpers and Splitters," 23.

53. David Ledbetter, interview by author, Danville, PA, March 21, 2012.

54. W. J. Fibison and B. S. Emanuel, "Molecular Mapping in DiGeorge Syndrome," *American Journal of Human Genetics* 41 (1987): A119.

55. Deborah Driscoll, interview by author, Philadelphia, November 29, 2011.

56. D. A. Driscoll, M. L. Budarf, and B. S. Emanuel, "A Genetic Etiology for DiGeorge Syndrome: Consistent Deletions and Microdeletions of 22q11," *American Journal of Human Genetics* 50, no. 5 (1992): 924–33.

57. Emanuel interview; D. A. Driscoll, N. B. Spinner, M. L. Budarf, D. M. McDonald-McGinn, E. H. Zackai, R. B. Goldberg, R. J. Shprintzen, H. M. Saal, J. Zonana, M. C. Jones, et al., "Deletions and Microdeletions of 22q11.2 in Velo-Cardio-Facial Syndrome," *American Journal of Medical Genetics* 44, no. 2 (1992): 261–68.

58. D. A. Driscoll, M. L. Budarf, and B. S. Emanuel, "Antenatal Diagnosis of DiGeorge Syndrome," *Lancet* 338, no. 8779 (1991): 1390–91, 1390; D. A. Driscoll, J. Salvin, B. Sellinger, M. L. Budarf, D. M. McDonald-McGinn, E. H. Zackai, and B. S. Emanuel, "Prevalence of 22q11 Microdeletions in DiGeorge and Velocardiofacial Syndromes: Implications for Genetic Counseling and Prenatal Diagnosis," *Journal of Medical Genetics* 30, no. 10 (1993): 813–17.

59. Driscoll et al., "Prevalence of 22q11 Microdeletions," 815.

60. R. J. Shprintzen, "Velocardiofacial Syndrome and DiGeorge Sequence," *Journal of Medical Genetics* 31, no. 5 (1994): 423–24, 424.

61. D. I. Wilson, J. Burn, P. Scambler, and J. Goodship, "DiGeorge Syndrome: Part of Catch 22," *Journal of Medical Genetics* 30, no. 10 (1993): 852–56.

62. Shprintzen, "Velocardiofacial Syndrome and DiGeorge Sequence," 424; McKusick, "On Lumpers and Splitters."

63. R. J. Shprintzen, "A History of the 'Discovery' of VCFS (or Whatever You Choose to Call It): Why We Have So Many Names for One Condition," *Velo-Cardio-Facial Syndrome Journal* 1, no. 1 (2013): 1–7, accessible at www.vcfscenter.com/Journal.html; Robert Shprintzen, telephone interview by author, March 26, 2015; Shprintzen's name, along with that of his colleague Rosalie Goldberg, was used in eponyms for two other disorders, Shprintzen-Goldberg and Goldberg-Shprintzen syndromes.

64. Shprintzen, "Velocardiofacial Syndrome and DiGeorge Sequence," 424; Shprintzen interview; R. Hirschhorn, "The Blind Men and the Rheumatoid Elephant," *New England Journal of Medicine* 293, no. 11 (1975): 554–55; J. P. Kassirer, "The Wild Goose Chase and the Elephant's Relevance," *Journal of the American Medical Association* 256, no. 2 (1986): 256–57; A. J. Tobin, "Molecular Biology and Schizophrenia: Lessons from Huntington's Disease," *Schizophrenia Bulletin* 13, no. 1 (1987): 199–203.

65. E. A. Wulfsberg, J. Leana-Cox, and N. Giovanni, "What's in a Name? Chromosome 22q Abnormalities and the DiGeorge, Velocardiofacial and Conotruncal Anomalies Face Syndromes," *American Journal of Medical Genetics* 65, no. 4 (1996): 317–19, 317–8; McKusick, "Human Genome through the Eyes of a Geneticist," 20.

66. Wulfsberg, Leana-Cox, and Giovanni, "What's in a Name?, 318; R. J. Shprintzen, "The Name Game," *Perspectives on Speech Science and Orofacial Disorders* 8 (1998): 7–11, 10. The use of decimal points in the microdeletion designations (e.g., 22q11.2) represented an increased level of band detection made possible by high-resolution cytogenetic techniques (see chapter 3).

67. D. M. McDonald-McGinn, E. H. Zackai, and D. Low, "What's in a Name? The 22q11.2 Deletion," *American Journal of Medical Genetics* 72, no. 2 (1997): 247.

68. D. M. McDonald-McGinn and E. H. Zackai, "The History of the 22q11.2 Deletion," *22q and You* newsletter (1996), accessible at www.cbil.upenn.edu/VCFS/22qandyou/history.html.

69. McKusick, "Human Genome through the Eyes of a Geneticist," 20.

70. R. J. Shprintzen and K. J. Golding-Kushner, *Velo-Cardio-Facial Syndrome* (San Diego: Plural, 2008), 1:16.

71. Hogan, "Medical Eponyms"; Daniel Navon, "Genomic Designation: How Genetics Can Delineate New, Phenotypically Diffuse Medical Categories," *Social Studies of Science* 20, no. 10 (2011): 1–24.

72. Shprintzen and Golding-Kushner, *Velo-Cardio-Facial Syndrome,* 16.

73. "Same Name Campaign," accessible at www.22q.org/index.php/foundation-information/same-name-campaign.

74. Shprintzen interview.

75. Ibid.; F. M. Mikhail, R. D. Burnside, B. Rush, J. Ibrahim, R. Godshalk, S. L. Rutledge, N. H. Robin, M. D. Descartes, and A. J. Carroll, "The Recurrent Distal 22q11.2 Microdeletions Are Often de novo and Do Not Represent a Single Clinical Entity: A Proposed Categorization System," *Genetic Medicine* 16, no. 1 (2014): 92–100.

76. Rabeharisoa et al., "From 'Politics of Numbers.'"

77. R. W. Marion, "A History of the 'Discovery' of VCFS (or Whatever You Choose to Call It): A Trainee's Perspective," *Velo-Cardio-Facial Syndrome Journal* 1, no. 1 (2013): 28, accessible at www.vcfscenter.com/Journal.html.

78. Rabinow, "Artificiality and Enlightenment"; Navon and Shwed, "Chromosome 22q11.2 Deletion"; Rabeharisoa et al., "From 'Politics of Numbers.'"

79. Navon, "Genomic Designation."

80. Soraya de Chadarevian and Harmke Kamminga, eds., *Molecularizing Biology and Medicine: New Practices and Alliances, 1910s–1970s* (Amsterdam: Harwood Academic, 1998).

81. Shprintzen interview.

82. Mikhail et al., "Recurrent Distal 22q11.2 Microdeletions"; McKusick, "Human Genome through the Eyes of a Geneticist," 20.

Chapter 6. Getting the Whole Picture

1. David Haussler, interview by author, Santa Cruz, CA, February 29, 2012.

2. Reed Pyeritz, interview by author, Philadelphia, April 18, 2012; Soraya de Chadarevian, "Mutations in the Nuclear Age," in *Making Mutations: Objects, Practices, Contexts, Preprint 393,* ed. Luis Campos and Alexander von Schwerin (Berlin: Max Plank Institute for the History of Science, 2010), 179–88, 180.

3. A. Kallioniemi, O. P. Kallioniemi, D. Sudar, D. Rutovitz, J. W. Gray, F. Waldman, and D. Pinkel, "Comparative Genomic Hybridization for Molecular Cytogenetic Analysis of Solid Tumors," *Science* 258, no. 5083 (1992): 818–21, 818.

4. Ibid.; B. J. Trask, "Human Cytogenetics: 46 Chromosomes, 46 Years and Counting," *Nature Reviews Genetics* 3, no. 10 (2002): 769–78.

5. A. Kallioniemi et al., "Comparative Genomic Hybridization."

6. O. P. Kallioniemi, A. Kallioniemi, D. Sudar, D. Rutovitz, J. W. Gray, F. Waldman, and D. Pinkel, "Comparative Genomic Hybridization: A Rapid New Method for Detecting and Mapping DNA Amplification in Tumors," *Seminars in Cancer Biology* 4, no. 1 (1993): 41–46, 44.

7. M. Bentz, K. Huck, S. du Manoir, S. Joos, C. A. Werner, K. Fischer, H. Dohner, and P. Lichter, "Comparative Genomic Hybridization in Chronic B-Cell Leukemias Shows a High Incidence of Chromosomal Gains and Losses," *Blood* 85, no. 12 (1995): 3610–18.

8. S. Solinas-Toldo, S. Lampel, S. Stilgenbauer, J. Nickolenko, A. Benner, H. Dohner, T. Cremer, and P. Lichter, "Matrix-Based Comparative Genomic Hybridization: Biochips to Screen for Genomic Imbalances," *Genes, Chromosomes & Cancer* 20, no. 4 (1997): 399–407.

9. A. C. Pease, D. Solas, E. J. Sullivan, M. T. Cronin, C. P. Holmes, and S. P. Fodor, "Light-Generated Oligonucleotide Arrays for Rapid DNA Sequence Analysis," *Proceedings of the National Academy of the Sciences* 91, no. 11 (1994): 5022–26.

10. M. Chee, R. Yang, E. Hubbell, A. Berno, X. C. Huang, D. Stern, J. Winkler, D. J. Lockhart, M. S. Morris, and S. P. Fodor, "Accessing Genetic Information with High-Density DNA Arrays," *Science* 274, no. 5287 (1996): 610–14.

11. Solinas-Toldo et al., "Matrix-Based Comparative Genomic Hybridization."

12. D. Pinkel, R. Segraves, D. Sudar, S. Clark, I. Poole, D. Kowbel, C. Collins, W. L. Kuo, C. Chen, Y. Zhai, S. H. Dairkee, B. M. Ljung, J. W. Gray, and D. G. Albertson, "High Resolution Analysis of DNA Copy Number Variation Using Comparative Genomic Hybridization to Microarrays," *Nature Genetics* 20, no. 2 (1998): 207–11.

13. Jean-Paul Gaudillière and Hans-Jorg Rheinberger, *From Molecular Genetics to Genomics: The Mapping Cultures of Twentieth-Century Genetics* (London: Routledge, 2004); V. G. Cheung, N. Nowak, W. Jang, I. R. Kirsch, S. Zhao, X.-N. Chen, T. S. Furey, U.-J. Kim, W.-L. Kuo, M. Oliver, et al., "Integration of Cytogenetic Landmarks into the Draft Sequence of the Human Genome," *Nature* 409, no. 6822 (2001): 953–58.

14. J. R. Korenberg, X. N. Chen, Z. Sun, Z. Y. Shi, S. Ma, E. Vataru, D. Yimlamai, J. S. Weissenbach, H. Shizuya, M. I. Simon, S. S. Gerety, H. Nguyen, I. S. Zemsteva, L. Hui, J. Silva, X. Wu, B. W. Birren, and T. J. Hudson, "Human Genome Anatomy: Bacs Integrating the Genetic and Cytogenetic Maps for Bridging Genome and Biomedicine," *Genome Research* 9, no. 10 (1999): 994–1001, 994.

15. Haussler interview; de Chadarevian, "Mutations in the Nuclear Age," 180.

16. T. G. Wolfsberg, K. A. Wetterstrand, M. S. Guyer, F. S. Collins, and A. D. Baxevanis, "A User's Guide to the Human Genome, Introduction: Putting It Together," *Nature Genetics* 32 (2002): 5–8.

17. Haussler interview.

18. Wolfsberg et al., "User's Guide to the Human Genome."

19. Haussler interview.

20. E. S. Lander, L. M. Linton, B. Birren, C. Nusbaum, M. C. Zody, J. Baldwin, K. Devon, K. Dewar, M. Doyle, and W. FitzHugh, "Initial Sequencing and Analysis of the Human Genome," *Nature* 409, no. 6822 (2001): 860–921.

21. Haussler interview.

22. Cheung et al., "Integration of Cytogenetic Landmarks."

23. Haussler interview.

24. Cheung et al., "Integration of Cytogenetic Landmarks," 954.

25. T. S. Furey and D. Haussler, "Integration of the Cytogenetic Map with the Draft Human Genome Sequence," *Human Molecular Genetics* 12, no. 9 (2003): 1037–44, 1037.

26. David Ledbetter, interview by author, Danville, PA, March 21, 2012.

27. Beverly Emanuel, interview by author, Philadelphia, November 9, 2011.

28. Kurt Hirschhorn, interview by author, New York, January 26, 2012.

29. Uta Francke, interview by author, Palo Alto, CA, February 27, 2012.

30. Deborah Driscoll, interview by author, Philadelphia, November 29, 2011.

31. Pyeritz interview.

32. Dorothy Warburton, telephone interview by author, May 11, 2011.

33. Ledbetter interview.

34. Robert Nicholls, telephone interview by author, April 5, 2012.

35. Driscoll interview.

36. L. G. Shaffer and B. A. Bejjani, "A Cytogeneticist's Perspective on Genomic Microarrays," *Human Reproductive Update* 10, no. 3 (2004): 221–26.

37. L. E. Vissers, B. B. de Vries, K. Osoegawa, I. M. Janssen, T. Feuth, C. O. Choy, H. Straatman, W. van der Vliet, E. H. Huys, A. van Rijk, D. Smeets, C. M. van Ravenswaaij-Arts, N. V. Knoers, I. van der Burgt, P. J. de Jong, H. G. Brunner, A. G. van Kessel, E. F. Schoenmakers, and J. A. Veltman, "Array-Based Comparative Genomic Hybridization for the Genomewide Detection of Submicroscopic Chromosomal Abnormalities," *American Journal of Human Genetics* 73, no. 6 (2003): 1261–70.

38. C. Shaw-Smith, R. Redon, L. Rickman, M. Rio, L. Willatt, H. Fiegler, H. Firth, D. Sanlaville, R. Winter, L. Colleaux, M. Bobrow, and N. P. Carter, "Microarray Based Comparative Genomic Hybridisation (Array-CGH) Detects Submicroscopic Chromosomal Deletions and Duplications in Patients with Learning Disability / Mental Retardation and Dysmorphic Features," *Journal of Medical Genetics* 41, no. 4 (2004): 241–48, 246.

39. Ibid.

40. Shaffer and Bejjani, "Cytogeneticist's Perspective on Genomic Microarrays," 222–23.

41. B. A. Bejjani, R. Saleki, B. C. Ballif, E. A. Rorem, K. Sundin, A. Theisen, C. D. Kashork, and L. G. Shaffer, "Use of Targeted Array-Based CGH for the Clinical Diagnosis of Chromosomal Imbalance: Is Less More?," *American Journal of Medical Genetics Part A* 134, no. 3 (2005): 259–67; A. S. Ishkanian, C. A. Malloff, S. K. Watson, R. J. DeLeeuw, B. Chi, B. P. Coe, A. Snijders, D. G. Albertson, D. Pinkel, M. A. Marra, V. Ling, C. MacAulay, and W. L. Lam, "A Tiling Resolution DNA Microarray with Complete Coverage of the Human Genome," *Nature Genetics* 36, no. 3 (2004): 299–303.

42. L. G. Shaffer, C. D. Kashork, R. Saleki, E. Rorem, K. Sundin, B. C. Ballif, and B. A. Bejjani, "Targeted Genomic Microarray Analysis for Identification of Chromosome Abnormalities in 1500 Consecutive Clinical Cases," *Journal of Pediatrics* 149, no. 1 (2006): 98–102.

43. L. Rickman, H. Fiegler, N. P. Carter, and M. Bobrow, "Prenatal Diagnosis by Array-CGH," *European Journal of Medical Genetics* 48, no. 3 (2005): 232–40.

44. Evelyne Shuster, "Microarray Genetic Screening: A Prenatal Roadblock for Life?," *Lancet* 369, no. 9560 (2007): 526–29, 527–28.

45. E. Check, "Fetal Genetic Testing: Screen Test," *Nature* 438, no. 7069 (2005): 733–34, 733.

46. T. Sahoo, S. W. Cheung, P. Ward, S. Darilek, A. Patel, D. del Gaudio, S. H. Kang, S. R. Lalani, J. Li, S. McAdoo, A. Burke, C. A. Shaw, P. Stankiewicz, A. C. Chinault, I. B. Van den Veyver, B. B. Roa, A. L. Beaudet, and C. M. Eng, "Prenatal Diagnosis of Chromosomal Abnormalities Using Array-Based Comparative Genomic Hybridization," *Genetics in Medicine* 8, no. 11 (2006): 719–27.

47. L. G. Shaffer, J. Coppinger, S. Alliman, B. A. Torchia, A. Theisen, B. C. Ballif, and B. A. Bejjani, "Comparison of Microarray-Based Detection Rates for Cytogenetic Abnormalities in Prenatal and Neonatal Specimens," *Prenatal Diagnosis* 28, no. 9 (2008): 789–95.

48. J. Coppinger, S. Alliman, A. N. Lamb, B. S. Torchia, B. A. Bejjani, and L. G. Shaffer, "Whole-Genome Microarray Analysis in Prenatal Specimens Identifies Clinically Significant Chromosome Alterations without Increase in Results of Unclear Significance Compared to Targeted Microarray," *Prenatal Diagnosis* 29, no. 12 (2009): 1156–66.

49. L. G. Shaffer, M. P. Dabell, A. J. Fisher, J. Coppinger, A. M. Bandholz, J. W. Ellison, J. B. Ravnan, B. S. Torchia, B. C. Ballif, and J. A. Rosenfeld, "Experience with Microarray-Based Comparative Genomic Hybridization for Prenatal Diagnosis in Over 5000 Pregnancies," *Prenatal Diagnosis* 32, no. 10 (2012): 976–85, 983.

50. A. Breman, A. N. Pursley, P. Hixson, W. Bi, P. Ward, C. A. Bacino, C. Shaw, J. R. Lupski, A. Beaudet, and A. Patel, "Prenatal Chromosomal Microarray Analysis in a Diagnostic Laboratory; Experience with > 1000 Cases and Review of the Literature," *Prenatal Diagnosis* 32, no. 4 (2012): 351–61, 358; Shaffer et al., "Experience with Microarray-Based Comparative."

51. R. J. Wapner, C. L. Martin, B. Levy, B. C. Ballif, C. M. Eng, J. M. Zachary, M. Savage, L. D. Platt, D. Saltzman, and W. A. Grobman, "Chromosomal Microarray versus Karyotyping for Prenatal Diagnosis," *New England Journal of Medicine* 367, no. 23 (2012): 2175–84, 2181.

52. B. A. Bernhardt, D. Soucier, K. Hanson, M. S. Savage, L. Jackson, and R. J. Wapner, "Women's Experiences Receiving Abnormal Prenatal Chromosomal Microarray Testing Results," *Genetics in Medicine* 15, no. 2 (2012): 139–45, 140–41; M. Reiff, B. A. Bernhardt, S. Mulchandani, D. Soucier, D. Cornell, R. E. Pyeritz, and N. B. Spinner, "'What Does It Mean?': Uncertainties in Understanding Results of Chromosomal Microarray Testing," *Genetics in Medicine* 14, no. 2 (2012): 250–58, 253.

53. Breman et al., "Prenatal Chromosomal Microarray Analysis," 358.

54. G. McGillivray, J. A. Rosenfeld, R. J. McKinlay Gardner, and L. H. Gillam, "Genetic Counselling and Ethical Issues with Chromosome Microarray Analysis in Prenatal Testing," *Prenatal Diagnosis* 32, no. 4 (2012): 389–95, 391.

55. Ledbetter interview.

56. M. C. de Wit, M. I. Srebniak, L. C. Govaerts, D. Van Opstal, R. J. Galjaard, and A. T. Go, "Additional Value of Prenatal Genomic Array Testing in Fetuses with Isolated Structural Ultrasound Abnormalities and a Normal Karyotype: A Systematic Review of the Literature," *Ultrasound, Obstetrics & Gynecology* 43, no. 2 (2014): 139–46, 144.

57. Antina de Jong, Wybo J. Dondorp, Merryn V. E. Macville, Christine E. M. de Die-Smulders, Jan M. M. van Lith, and Guido M. W. R. de Wert, "Microarrays as a Diagnostic Tool in Prenatal Screening Strategies: Ethical Reflection," *Human Genetics* 133, no. 2 (2014): 163–72, 168–69.

58. M. Allyse, M. A. Minear, E. Berson, S. Sridhar, M. Rote, A. Hung, and S. Chandrasekharan, "Non-Invasive Prenatal Testing: A Review of International Implementation and Challenges," *International Journal of Womens Health* 7 (2015): 113–26.

59. American College of Obstetricians and Gynecologists, "ACOG Committee Opinion No. 581: The Use of Chromosomal Microarray Analysis in Prenatal Diagnosis," *Obstetrics and Gynecology* 122 (2013): 1374–77, 1376.

60. Ledbetter interview; J. J. Yunis and M. E. Chandler, "High-Resolution Chromosome Analysis in Clinical Medicine," *Progress in Clinical Pathology* 7 (1977): 267–88.

61. Advertisement for Prenatal Scan Array CGH Test by Combimatrix Molecular Diagnostics. This advertisement was aimed at prenatal diagnostic providers and given out by Combimatrix representatives at a Prenatal Diagnosis conference in Philadelphia in May 2011. The advertisement features an image of a pregnant belly with a puzzle piece containing an image of results from microarray testing. While primarily targeted toward providers, the advertisement's imagery and language may also be meant to appeal to patients as part of medical consultation.

62. Bernhardt et al., "Women's Experiences"; Reiff et al., "'What Does It Mean?'"

Epilogue. The Genomic Gaze

1. Donna J. Haraway, *Simians, Cyborgs, and Women: The Reinvention of Nature* (New York: Routledge, 1990), 194.

2. Adele E. Clarke, Janet K. Shim, Laura Mamo, Jennifer Ruth Fosket, and Jennifer R. Fishman, "Biomedicalization: Technoscientific Transformations of Health, Illness, and U.S. Biomedicine," *American Sociological Review* 68, no. 2 (2003): 161–94; Adele E. Clarke, Laura Mamo, Jennifer Ruth Fosket, Jennifer R. Fishman, and Janet K. Shim, eds., *Biomedicalization: Technoscience, Health, and Illness in the U.S.* (Durham, NC: Duke University Press, 2010); Nikolas Rose, *The Politics of Life Itself: Biomedicine, Power, and Subjectivity in the Twenty-First Century* (Princeton, NJ: Princeton University Press, 2007), 12.

3. Katie Featherstone and Paul Atkinson, *Creating Conditions: The Making and Remaking of a Genetic Syndrome* (London: Routledge, 2012), 113; Katie Featherstone, J. Latimer, P. Atkinson, D. T. Pilz, and A. Clarke, "Dysmorphology and the Spectacle of the Clinic," *Sociology of Health & Illness* 27, no. 5 (2005): 551–74; Alison Shaw, "Interpreting Images: Diagnostic Skill in the Genetics Clinic," *Journal of the Royal Anthropological Institute* 9, no. 1 (2003): 39–55.

4. Joanna Latimer, *The Gene, the Clinic, and the Family: Diagnosing Dysmorphology, Reviving Medical Dominance* (New York: Routledge, 2013); Joanna Latimer, Katie Featherstone, Paul Atkinson, Angus Clarke, Daniela T. Pilz, and Alison Shaw, "Rebirthing the Clinic: The Interaction of Clinical Judgment and Genetic Technology in the Production of Medical Science," *Science, Technology & Human Values* 31, no. 5 (2006): 599–630.

5. Vololona Rabeharisoa and Pascale Bourret, "Staging and Weighting Evidence in Biomedicine," *Social Studies of Science* 39, no. 5 (2009): 691–715, 704; Soraya de Chadarevian, *Designs for Life: Molecular Biology after World War II* (Cambridge: Cambridge University Press, 2002); Ilana Löwy, *Between Bench and Bedside: Science, Healing, and Interleukin-2 in a Cancer Ward* (Cambridge, MA: Harvard University Press, 1996).

6. Peter Keating and Alberto Cambrosio, *Biomedical Platforms: Realigning the Normal and the Pathological in Late-Twentieth Century Medicine* (Cambridge, MA: MIT Press, 2003); Peter Keating and Alberto Cambrosio, "Does Biomedicine Entail the Successful Reduction of Pathology to Biology?," *Perspectives in Biology and Medicine* 47, no. 3 (2004): 357–71.

7. Soraya de Chadarevian, "Mutations in the Nuclear Age," in *Making Mutations: Objects, Practices, Contexts, Preprint 393,* ed. Luis Campos and Alexander von Schwerin (Berlin: Max Plank Institute for the History of Science, 2010), 179–88, 180.

Index